D0041914

MAPPING THE COSMOS

NEW SERIES, NO. 4

Mapping the Cosmos

EDITED BY JANE CHANCE

AND R. O. WELLS, JR.

Rice University Press   Houston, Texas

© 1985 by Rice University

All rights reserved

Printed in the United States of America

First Edition, 1985

Requests for permission to reproduce material

from this work should be addressed to:

Rice University Press

Rice University

Post Office Box 1892

Houston, Texas 77251

"The Influence of Alchemy on Newton," by Richard Westfall, is reprinted by
permission of Wilfrid Laurier University Press.

Plate 1 from Ms. 48 of the *Roman de la Rose* is used
with permission of the Pierpont Morgan Library.
Plates 2 and 3, Wilton Diptych, are used by courtesy
of the trustees of the National Gallery, London.
Plate 7, from the *Liber Floridus* of Lambert Ms. 724/1596 in the Musée Condé,
Chantilly, is used by permission of Giraudon/Art Resource, New York.

Library of Congress Cataloging in Publication Data

Main entry under title:

Mapping the cosmos.

(Rice University Press; new. ser. no. 4)

Includes index.

1. Science—History.   2. Knowledge, Theory of—History.

I. Chance, Jane, 1945– .   II. Wells, R. O. (Raymond

O'Neill), 1940– .   III. Series.

AS36.W65 n.s., no. 4 [Q125] 080 s [509]   85-2219

Library of Congress Catalog Card Number 84-62986

ISBN 0-89263-258-5

To the memory of Salomon Bochner

CONTENTS

SEVEN

The New Science of Information
Howard L. Resnikoff

*The ornament used at the beginning of each
chapter is an ancient symbol that represents the
universe as it was once conceived. The dark
center is the orb of earth and water. The inner
ring is the aerial ocean. The outer ring is the
highest heaven. From Rudolph Koch,* The Book
of Signs.

Introduction

Jane Chance and R. O. Wells, Jr.

From the mythological world of the Homeric epic to the very recent developments in the modern science of information theory, we see how the perception of the world around us and of humanity's place in this world has evolved and changed, often dramatically, over the centuries. While there is no modern word that describes precisely the human attempt to map the cosmos, there does exist an ancient Latin word for that venerable concept of knowledge that includes but also transcends what we think of as science.

The modern word "science" often is understood to refer to systematized knowledge of nature and the physical world rather than its more general denotation as knowledge itself as opposed to intuition or belief. And yet the Latin word from which it derives, *scientia*, means exactly that: "knowledge." *Scientia* embraces many different kinds of knowledge—from ancient sciences like geometry, astronomy, physiognomy, and alchemy, and ancient "arts" like grammar, rhetoric, and logic, to more modern sciences like biology, the behaviorial sciences, and the new technology of computer science. In this volume of essays by contemporary scholars of diverse backgrounds, we see that the history of ideas—the evolution of *scientia*—provides a means of linking ideas in the contemporary fields of the sciences and the humanities.

In the Middle Ages the different branches of knowledge were related; philosophy was understood to encompass all knowledge, including science. Study traditionally began with grammar, rhetoric, and dialectic and progressed to the *quadrivium*, the other four of the liberal arts:

arithmetic, geometry, music, and astronomy. After completing these fields, the student arrived at philosophy, for which the seven liberal arts were handmaidens, and finally one used all of these branches of knowledge in the study of the Bible. Indeed, Martianus Capella in the fifth century wrote a very strange handbook on the seven liberal arts in which the preface consisted of an allegory of the wedding of eloquence (Mercury) and wisdom (Philology), attended by the Seven Liberal Arts themselves. The wedding was intended to stress the goal of education as the combination of the divine and the human. Definitions of knowledge often became quite complex, while always maintaining the ideal of the relatedness of *scientia*. Hugh of St. Victor, for example, in his twelfth-century work *Didascalicon*, declared that "philosophy is divided into theoretical, practical, mechanical, and logical. These four contain all knowledge [*scientia*]." Hugh's holistic definition of *scientia* may surprise. He divides the theoretical into theology, physics, and mathematics, the latter of which embraced the four arts of the *quadrivium:* arithmetic, geometry, astronomy, and music. He divides the practical into solitary, private, and public, by which he means ethics—how one should live one's life, whether alone, in a family, or in a society. He divides the mechanical into fabric-making, armament, commerce, agriculture, hunting, medicine, and theatrics. Finally, he divides the logical into grammar and argument, or the *trivium:* grammar, rhetoric, and dialectic.

The sequence of ideas discussed in the following essays represents an ancient tradition based on Hugh's definition of *scientia* as a linked hierarchy of all knowledge. Such knowledge, in the sense of the German word *Wissenschaft*, was the ideal of the renowned mathematician Salomon Bochner, who founded an institute at Rice University in 1981 that was intended to bring together individuals who had in common an interest in the history of ideas in science and culture or an interest in making connections between two different branches of knowledge or disciplines. Professor Bochner well understood both the German concept of *Wissenschaft* and the medieval definition of *scientia* as the Latin word for knowledge; hence the institute was named "Scientia." Many of the essays in this volume were first delivered as Scientia colloquia at Rice University.

In each essay the change in a particular idea is explored and explained within one of these ancient or modern "sciences." The impact of antiquity on the particular science is traced, and each essay ultimately demonstrates that Ecclesiastian adage that there is "nothing new under the sun." The volume thus shows the ways in which our knowledge of the universe and our attempt to offer something unique to that store of information become part of the larger pattern of history, and in particular the history of ideas.

The lectures as they were first conceived were individual attempts to explain a particular set of ideas to a nonspecialized audience. The resulting essays overlap remarkably well in their intellectual concerns. For instance, a specific piece of literature, Dante's *Commedia,* is used in one essay to explain the origin of mythography in ancient and medieval times and again in another to convey the ways in which medieval writers understood mathematical astronomy. In other essays towering thinkers like Ptolemy and Newton play different but distinctive roles in relation to mathematics and astronomy or in relation to history and alchemy. The unity of culture and the impact of specific individuals and their ideas on different aspects of human cultural and intellectual heritage become at times startlingly vivid when seen from different vantage points.

The first essay, by R. O. Wells, Jr., describes the evolution of the concepts of geometry and the ways in which various notions of geometry have been used to describe the perceivable universe. This perception varied greatly with time and with human knowledge of the extent and nature of the universe and has continued to expand dramatically to this day. The current "big bang theory" stretches contemporary incredulity as much as the more ancient theories of the zodiac or the epicycles of Ptolemy.

The mythic view of the cosmos is viewed through the lens of the literary traditions of ancient and medieval times in the essay by Jane Chance. Literature has always constituted a means of communicating contemporary perspectives of the world, both real and imagined. In this essay we discover the different levels of interpretation of the great epic poetry of the various civilizations represented: from the literal (and fictional) through the religious to the behavioral nature of humankind. Here the inner structure of society mirrors a cosmos filled with mythological figures—what must have represented, for the early poets, a form of what we would call "science fiction."

The essay by Albert Van Helden discusses cosmography in the late Middle Ages, in particular the ways in which the Ptolemaic system of cosmography entered western Europe through the astronomical literature of the Moslems. The focus on the evolution of mathematical astronomy from Ptolemy to Moslem thinkers reveals the extent to which medieval literary figures, dominated as they might have been by theological concerns, were aware of the latest scientific descriptions of the universe. This perception occurred at a very sophisticated level, with specific references in major works of literature to astronomical distances between the heavenly bodies. The medieval scholars were much less ignorant of their cosmological heritage than contemporary scholarship has credited them.

The essays by both Alfred David and Richard Westfall look at the interaction of cultural values with society's perceived notions about two very specific "sciences": physiognomy and alchemy. David examines the role of physical types (in particular, noses) in cultural value systems. Westfall depicts the seeming anomaly that one of the greatest natural scientists of all time, Sir Isaac Newton, devoted twenty-five years of effort and more than one million written words to the subject of alchemy, considered an unacceptable field of endeavor by the scientific community of his day.

The theme of what constitutes "proper" and "improper" science forms the basis of the essay by George Terrell, in which he examines the parallel developments in the behavioral and biological sciences in the late nineteenth century. The two disciplines used different paradigms but the same vocabulary, which often yielded conflicting results. There is a strong resemblance to the relationship between the alchemists and the scientific establishment of the earlier days, with evidence of religious fervor on both sides.

The last essay, by Howard Resnikoff, is of a slightly different character in that it gives us a glimpse of what will undoubtedly be maps of the future. Just as geometry has given us a picture of the universe around us— the nature of the universe being *information* in a codifiable form—so also has information technology. The exchange and transfer of information obeys laws as consistent and legitimate as the motion of particles behaving according to Newton's laws. Our generation is being exposed to a revolution in human thought and behavior that will probably be as dramatic as Newton's great contributions of three centuries ago. It is hoped that this collection of essays will broaden our perspective when we deal with future ideas of Homo sapiens' place in the cosmos.

We are grateful to the provost and academic vice-president of Rice University, William E. Gordon, for his continued support of Scientia and its colloquia series. We would also like to thank the dean of humanities, Allan J. Matusow, for supplying funds for the index. And finally, we appreciate the invaluable assistance of Ted Reed and Gayle Furlong.

O N E

Geometry and the Universe

R. O. Wells, Jr.

Geometry has been a part of human thought since the beginnings of civilization, and it certainly predates writing in the form of geometric figures preserved from prehistoric times. One has often used geometry and its symbolic representation to describe many different aspects of the real world surrounding the curious observer. In this essay I want to discuss how the concept of geometry has evolved through the centuries, starting with the ancient Greeks and moving up to some of the most recent achievements of contemporary mathematicians. At the same time, I will attempt to illustrate in what fashion the notions from geometry have been used by physical scientists to represent the universe around us. The notion of universe has evolved dramatically over the centuries. In particular, the concept of the largest and smallest objects in the universe seems to be changing drastically with time, especially in this rapidly evolving twentieth century. Geometric ideas have played an important role in all of these advances in our understanding of the cosmos. My objective here is to give you some idea of what this understanding is like. We will start with the zodiac of the Babylonians and proceed to the twentieth century's highly evolved quantum theory and relativity theory.

The Geometry of Euclid and Its Evolution through the Eighteenth Century

Euclid's work on the foundations of geometry, which was written in the third century B.C., has had a profound influence on civilization, comparable to very few other written documents. Most people today are familiar with some of the notions of Euclidean geometry as it is introduced in high school geometry courses around the country, but far fewer people are aware of the historic significance and the nature of this remarkable publication. Euclid's book was about the ideas of geometry in the plane and three-dimensional space. It represented the culmination of more than four hundred years of research by Greek mathematicians who, over the period of time from 600 B.C. to 200 B.C., slowly evolved the major ideas that were summarized and synthesized in the writings of Euclid. Out of the attempts and the need to use geometric ideas in everyday commerce, especially concerning surveying land and constructing buildings and other objects, evolved fundamental questions about the nature of length, area, volume, and angle. These are descriptive words that represent fundamental geometric concepts. They are geometric in that we can visualize in a pictorial fashion what these words represent.

The Greeks also introduced the notion of trying to reduce the study of a complicated object to its simplest components. Just as in writing we use letters, then words, then sentences, then books, then libraries to represent more and more complicated modes of thought, the Greek geometers wanted to reduce complicated geometric problems to the study of more elementary ones. For instance, one has the notion of square, triangle, and circle being fundamental and elementary geometric shapes. They extended this list of objects by using a three-dimensional point of view and added the conic sections to this list, which were defined as the intersection of a cone (generated by circles and straight lines in 3-space) with a plane. These conic sections were thus derivable from the more elementary concepts and include the familiar notions of parabola, ellipse, and hyperbola. These objects were the subject of much study, which culminated in a famous book by Apollonius entitled *Conic Sections*, which was later translated and played a pivotal role in the transition from the Greeks' mathematics to more modern mathematics, as we will mention again later in our discussion.

A fundamental accomplishment of the geometers of Greece was the introduction of the notion of logical deduction into the world of geometry. This was initiated by Thales of Miletus in ca. 600 B.C. by introducing the concept of a mathematical system with a set of axioms (supposed to be self-evident initially) from which all other true statements in the mathematical system could be derived. These derivations from the axioms used the laws of logic (a formal extension of old-fashioned common sense), and this became a model for all later mathematical disciplines, including some, like number theory, that have (at least on the surface) very little to do with geometry. In fact, the Greek word for what we nowadays call a mathematician was "geometer."

In the seventeenth century there was a major development in the linking up of algebra and geometry. This was the accomplishment of two contemporaries in France: Pierre Fermat (1601–1665) and René Descartes (1596–1650). Algebra had a tradition distinct from that of geometry. Algebra evolved from the attempt to "solve equations" that arose from various parts of the real world. For instance, one could say: given a rectangle whose area is 20 and whose length is 5, what is its width? Here we are trying to "solve for the width," given the details of the problem. The Babylonians (2000 B.C.–300 B.C.) gave recipes for solving this problem (answer: divide area by length to get width!), as well as much more complicated ones. They did not have a symbolic notation; they expressed their formulas in terms of words, just as we did above. They could solve linear equations, quadratic equations, and systems of equations in this manner. Many of the formulas we learn about in high school today, for instance the quadratic formula (see Fig. 1), were known (although not in our modern symbolic form) to the Babylonians roughly sixteen hundred years before Euclid wrote his treatise on geometry. The notation we are using here developed in the period roughly from 1200 A.D. to 1600 A.D.

Figure 1. *Quadratic formula*

$$x = \frac{-b \pm \sqrt{b^2 - 4ac}}{2a}$$

is the solution to the equation

$$ax^2 + bx + c = 0$$

The important discovery of Fermat and Descartes was that a geometric object (such as a circle, a straight line, or a conic section) could be represented as solutions of specified kinds of equations. This was achieved by means of a coordinate system (using Descartes's notation) where a point P in the plane is specified by giving coordinates (x,y), an ordered pair of numbers. We are familiar with coordinate systems in everyday life (latitude 15° N, 30° W; 10 miles north and then 15 miles west of a given point). The labeling on a sheet of graph paper represents explicitly a given coordinate system. We have, then, examples of solutions of equations corresponding to some familiar and some not-so-familiar geometric objects.

Figure 2. *Cartesian coordinates*

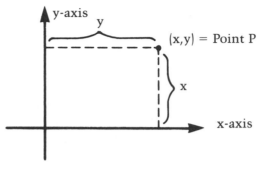

By using the symbolic notation for algebraic functions involving algebraic operations (Fig. 3), or by using labels for calculable nonalgebraic functions (Fig. 4), we have a relation between geometry and functions that can be manipulated according to mathematical laws. This manipulation gives a corresponding change in the geometric pattern associated with it. For instance, if we consider the sine curve (Fig. 4) and simply change x to $2x$ in the formula $y = \sin x$, obtaining $y = \sin 2x$, we get a new curve that has twice the frequency of the original curve (Fig. 5). On the other hand, if we multiply $\sin x$ by 2, obtaining $y = 2\sin x$, we have the same frequency, but with twice the magnitude (Fig. 5).

Thus, if we use geometric figures to represent real-world activity, we can use the associated symbolic notation to calculate various quantities associated with the geometric figures. For instance, one could ask for the area inside a curve defined by a certain equation in terms of the defining functions. In fact, this was one of the principal problems of calculus as it was developed in the seventeenth century.

Classical Models of the Universe: From the Babylonian Zodiac to Newton's Law of Universal Gravitation

A "model of the universe" is a description of nature of a certain kind. Namely, a model has a predictive ability built into it. If one focuses on the part of the universe described by the model, one should be able to predict future events. The first models of the universe were astronomical in nature, attempting to predict the future motions of stars, planets, the sun, and the moon. Today these models have evolved into what are now called "cosmological models" of the entire universe. The first of these models of which we have any extensive documentary evidence is

Figure 3. *Algebraic curves*

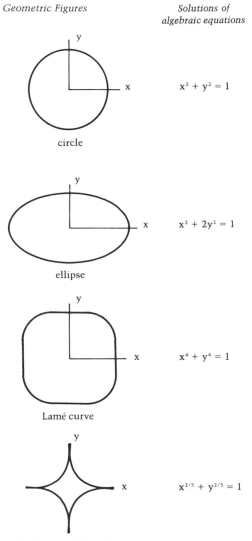

Geometric Figures

Solutions of algebraic equations

circle

$x^2 + y^2 = 1$

ellipse

$x^2 + 2y^2 = 1$

Lamé curve

$x^4 + y^4 = 1$

Lamé curve with vertices

$x^{2/5} + y^{2/5} = 1$

the zodiac of the Babylonians, utilized extensively during their flourishing period (2000 B.C. – 300 B.C.).

The Babylonians thought the earth was flat and the zodiac was a cylindrical strip of firmament that rotated overhead carrying with it stars and other heavenly bodies, including the sun and moon. The most in-

Figure 4. *Nonalgebraic curves*

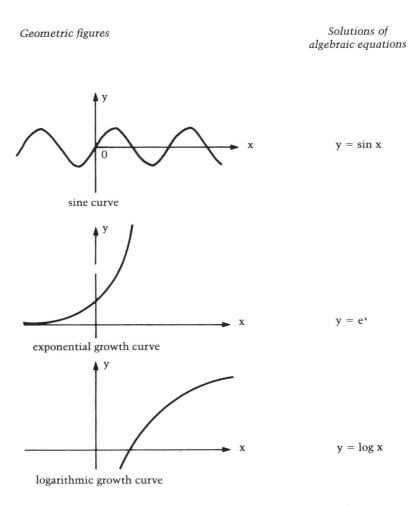

Geometric figures

*Solutions of
algebraic equations*

sine curve

$y = \sin x$

exponential growth curve

$y = e^x$

logarithmic growth curve

$y = \log x$

triguing aspect of their astronomical world was the notion of an eclipse. On the basis of *later* models we know that, for instance, a lunar eclipse is caused by the passing of the earth's shadow from the light of the sun over the moon. This requires the notion of two distinct round bodies (the earth and sun) with a physical light source geometrically on the other side (see Fig. 6). This picture is hard to visualize if one supposes that the sun and moon are both plastered to a cylindrical sheet as visualized in Figure 7, which was the model used by the Babylonians (and is the origin of contemporary astrology; in those days there was no distinction between astronomy and astrology). On the basis of many observa-

Figure 5. *Different sine curves*

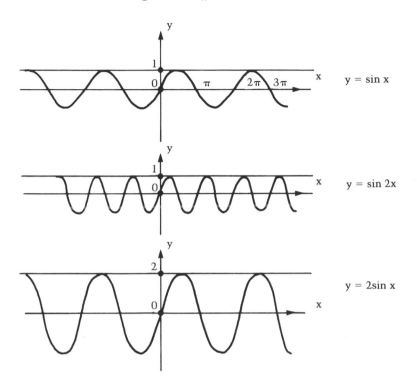

Figure 6. *Eclipse of the moon*

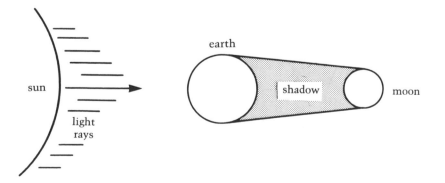

tions over the centuries, the Babylonians developed a remarkable ability to *predict* eclipses (including the percentage of the surface area eclipsed; not all eclipses were total). This they achieved by means of certain periodic functions of time (called "zig-zag functions" by contemporary historians), which graphically looked like the one illustrated in Figure 8.

The Babylonians had no such graphic representation. Their functions are represented by a table of values for which they had algorithmic rules for calculating. They were not tables of measurements, since many of the values were not physical observables. Only for certain ranges of the values could they obtain measurable displacements from known positions of various astronomical bodies. When certain values of the functions were achieved, one had an eclipse; and since the values are a function of time, one could predict the future because one could compute the values of the functions. These zig-zag functions predicted eclipses *without* having a geometric model of what an eclipse was. The later geometric models of eclipses of our solar system allowed one to construct such periodic functions as accurately as desired (depending on one's computational resources). Since the Babylonians did not have a geometric model, they could not necessarily refine their functions and make them more accurate because the only justification they had for their tables of functions was comparison with the actual eclipses. The models of Ptolemy, Copernicus, Kepler, and Newton each gave, in turn, a more refined geometric description of the solar system (and of much of the universe as understood by their contemporaries). In these models one had a much clearer concept of the rotation of the heavenly bodies around one another in some fashion, and it made sense to occasionally have the sun, earth, and moon juxtaposed on a straight line as illustrated in Figure 6. This was at the time the geometric understanding of the notion of an eclipse. The ability to predict the eclipses was then related to the ability to predict the relative motion of planetary bodies in general, and this was one of the major goals of the evolving modes of the solar system over the centuries.

Ptolemy wanted to reduce everything to the simplest geometric figure, the circle (or, in three-dimensional space, the sphere). His model of the universe had fixed stars on an outside spherical shell, the earth at the center, and a concentric sphere on which the sun and moon "rode." On the sphere for the sun, for instance, the sun was attached to a smaller sphere called an epicycle, whose center rotated with time on the bigger sphere, and the point that represented the center of the sun then moved in a complicated pattern due to the simultaneous motion of the two spheres (see Fig. 9). One can envision a gear box with connected gears or a bicycle tire rotating in the air with a small wheel attached to its rim. The various choices of radii and speeds of rotation could be

Figure 7. *Cylindrical firmament*

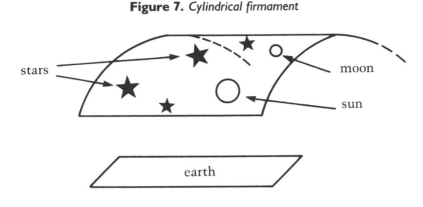

Figure 8. *Zig-zag functions of Babylonians*

matched to the planetary, solar, and lunar motions about the fixed earth. This was a complicated but quite successful model, which endured for about twelve hundred years. Copernicus (1473–1543) did not change the model at all, except that he interchanged the role of earth and sun, putting the sun at the center and the earth on a revolving epicycle on a concentric sphere along with the other planets. This simplified some of the parameters in the epicycle description, but the fundamental geometric picture was the same. Of course, the fundamental philosophical picture was not the same at all, since the earth was no longer the "center of the universe."

Kepler, on the other hand, proposed a more general set of laws from which one concluded that the planets moved about a (supposedly fixed) sun in elliptic orbits. Recalling that an ellipse is one of the fundamental geometric figures studied by Ptolemy's predecessor Apollonius, one can only wonder that Greek astronomers did not admit an ellipse as a possible planetary path. In some sense, it was not perfect enough for them.

One interesting footnote to this discussion is the fact that Copernicus noted in his working journal that the planetary orbits did indeed re-

Figure 9. *Epicycles of Ptolemy*

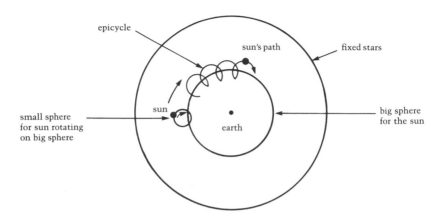

semble ellipses. However, he crossed out the page and went on to further epicyclic considerations, leaving this fundamental observation to Kepler. One of the keys to a better model is better information or experimental evidence. The astronomical observatory of Tycho Brahe (1546–1601) gave Kepler far more data than his predecessors with which to work. Thus, he could compare his new model with the earlier models in a convincing fashion. Without much data one is often reduced to authority, and in this case the authority was certainly on the Ptolemaic side. If there is little data, then many models might fit the data. On the other hand, if there is a lot of data, then one or two models will usually sweep the others aside, as they will agree convincingly with the evidence (until they too break down and it is clear that a completely new model is needed, as was the case with the onset of the nuclear age and the discovery of quantum mechanics). Newton's laws of gravitation are an example of a model that swept the competitors far from further consideration.

On the other hand, the geometric notion of a celestial sphere with fixed stars at a fixed radius from the earth (a model attributed to Ptolemy and his coworkers in Alexandria) is still very useful as a navigational model. We know the model is incorrect, but since it is still useful, we go right ahead and use it! For purposes of navigation on the surface of the earth, this particular geometric model (although theoretically completely wrong) is used extensively today by sailors and navigators around the world.

What is the major difference between Newton's law of gravitation and the Babylonian model of an eclipse? The answer is its *universality.* Newton prescribed that any two bodies were attracted to each other by a gravitational force that is inversely proportional to the inverse square of

Figure 10. *Celestial sphere*

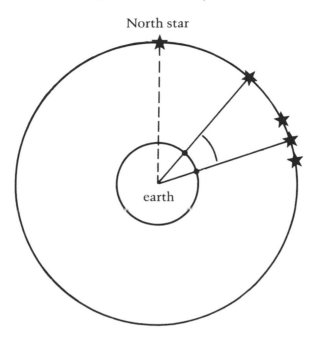

North star

earth

the distance between them. The proportionality involved the mass of each object (m_1 and m_2; see Fig. 11) as well as a universal constant G, independent of which bodies were involved. The mass is a measure of how much matter was in each object. If m_2 were the earth and m_1 referred to an object (such as a person) near the surface of the earth, then m_1 would correspond approximately to the usual notion of *weight* on the surface of the earth.

This law, when applied to larger systems of bodies, describes the solar system accurately enough to predict extremely well the observed motions of the planets as well as the position of rockets, satellites, and people in space in general. In particular, it predicts the eclipses we discussed earlier. The Babylonian model, with its zig-zag functions, predicted precisely eclipses but not much more. We would say that Newton's understanding was much deeper, and it is also quite clearly geometric in nature. We will see later how Einstein went one step further in providing an even deeper geometric understanding of the cosmos. But how did Newton's laws compare to, say, Kepler's laws? Kepler's laws were magnificent in their description of the solar system, and they made the Ptolemaic epicycle system obsolete. Today we still use Kepler's laws in our description of the solar system. But Newton's laws (from which Kepler's

Figure 11. *Newton's law of universal gravitation*

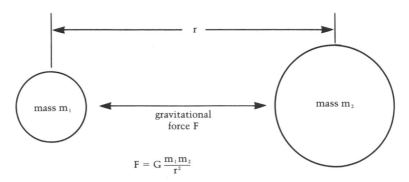

$$F = G \frac{m_1 m_2}{r^2}$$

laws can be derived) predicted the gravitational attraction of any two massive objects and, in particular, applied to masses of varying density. A modern post–World War II invention is the gravity-meter, which can be lowered into test holes drilled into the earth to measure the varying density in the crust of the earth. This has been very useful for detecting anomalies in the crust, which could contain petroleum deposits, for instance. This description of a structural variation of the earth's crust has nothing to do with the geometric structure of the solar system, and yet it is amenable to analysis by Newton's law of gravitation. Thus, we conclude that Newton's law is much deeper than Kepler's three laws (which applied only to the solar system), which, in turn, are much deeper than the Babylonian zig-zag functions (which applied only to eclipses).

In developing his laws of motion Newton invented (simultaneously with Leibniz in Germany) the calculus. This is the analytic study of rates of change of functions and of finding areas defined by the graphs of functions. It contrasted vividly with the Greeks' more "constant" or "uniform" geometry, as depicted in Figure 12, where we see contrasted the constant rate of change ("proportion") of the Greeks with the variable rate of change of Newton and Leibniz. Similarly, one finds a comparison of calculating areas (and also volumes) of simple versus more complicated geometric figures.

The calculus is an analytic tool for solving problems arising in geometry (and elsewhere), which had profound effects on civilization. The power of Newton's laws was their predictive ability. This power depended on the ability to *compute* the predicted behavior. For this the calculus and its progeny (for example, the theory of *differential equations*, a successor to the theory of algebraic equations of the Babylonians and the differential calculus of Newton and Leibniz) played an important role. The geometric model and the analytic ability to compute with the model have to go hand in hand.

Figure 12. *Comparison of calculus with earlier Greek analysis*

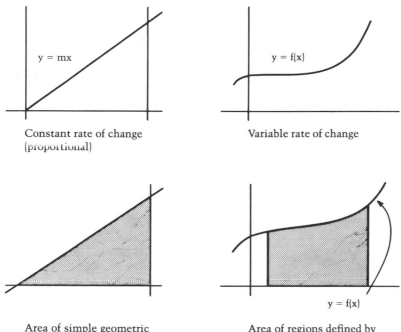

Greek viewpoint

Viewpoint of the calculus

y = mx

y = f(x)

Constant rate of change
(proportional)

Variable rate of change

Area of simple geometric
figures

Area of regions defined by
variable functions f(x)

y = f(x)

The Revolution of Geometric Thought in the Nineteenth Century and Its Modern Developments

After the discovery of the relationship between algebra and geometry in the early 1600s by Descartes and Fermat, geometry flourished, and the geometry of Euclid was amplified by the study of more complicated equations than those of the second order (conic sections) as well as by the discovery and applications of calculus discussed briefly in the previous section. This development proceeded unabated for some two hundred years, but always the fundamental background was that of Euclid. Thus, curves and surfaces of much more general type in the Euclidean 3-space, described by coordinates *(x,y,z)*, could be described and analyzed. One studied *arc length* (length of a given curve), *surface area,* and *curvature* (measure of deviation from being a straight line or flat plane)

of geometric objects. Essentially many of the classical problems of geometry were much more accessible, and one used the curves and surfaces to represent solutions of physical problems. The curves and surfaces studied had a behavior that was not necessarily uniform (like in a
circle or a straight line), and the calculus, which was invented by Leibniz and Newton for studying variable rates of change, turned out to be
ideal for studying the behavior of these geometric objects. On the other
hand, the fundamental notion of distance between points had not varied
since the time of the Greeks. This was based on the famous theorem of
Pythagoras (see Fig. 15). The most fundamental objects were still straight
lines, and the shortest distance between two points was the length of the
straight line segment joining them. The axioms of Greek geometry were
still the governing principles, even though algebra and calculus had
changed the geometric landscape considerably.

In the nineteenth century, some 2,100 years after the appearance of
Euclid's *Elements*, several independent mathematicians put forth ideas
that were very radical at the time, and the set of ideas goes today by the
name of *nonEuclidean geometry*. The very name indicates the "heresy"
of the ideas, just as the word "protestant" referred to the protest movement against the Catholic Church several centuries earlier. The mathematicians were Bolyai, Lobashewski, and Gauss in the first half of the
nineteenth century, but it took decades for the work to be accepted, and
only in the twentieth century has the full significance of their ideas be-

Figure 13. *Euclidian 3-space with curves and surfaces*

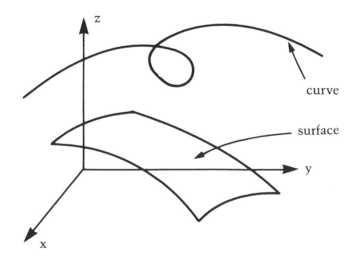

Figure 14. *Curves in 3-space appearing in physical problems*

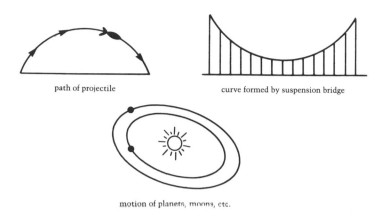

path of projectile   curve formed by suspension bridge

motion of planets, moons, etc.

Figure 15. *Euclidian distance between two points and Pythagoras' theorem*

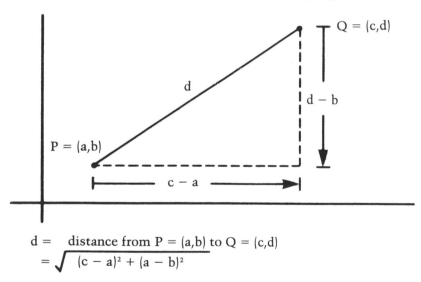

$$d = \text{distance from } P = (a,b) \text{ to } Q = (c,d)$$
$$= \sqrt{(c - a)^2 + (a - b)^2}$$

gun to penetrate modern science. The simplest example of a nonEuclidean geometry (which was surprisingly not the first) is the spherical geometry of Riemann, in which one views a sphere as a geometric space.
The "straight lines" in this space are the great circles, which are the
paths on the sphere that join two points and have the shortest length on
the sphere. The modern traveler is familiar with great circle routes of

the airlines. The axioms of nonEuclidean geometries are similar to but not quite the same as those of Euclidean geometry. In the beginning it was hard to believe there was any other than the revered geometry of Euclid. But gradually there came acceptance of *different* axiomatic mathematical systems, not only in geometry, but also in algebra, where Hamilton invented his *quaternions* (in 1843), a system of "numbers" in which $a \cdot b \neq b \cdot a$. This was revolutionary at the time, but now we have many mathematical systems that do not satisfy the commutative law $a \cdot b = b \cdot a$. In earlier times we did not give a name to something self-evident and for which there was no contrasting property. The term *Euclidean geometry* really did not exist until there was an antithesis against which it could be compared. Today we teach the "commutative properties of real numbers" to elementary school students. Before, we would merely have described the numbers and their "natural and self-evident" properties.

There were two other major discoveries in the nineteenth century that are related to the nonEuclidean geometry above and, in fact, can be considered as generalizations of same. Namely, Gauss discovered in 1827 the notion of *intrinsic geometry*, and Riemann formulated the notion of a *manifold* in 1854. Both of these concepts were new to the world of geometry and added immensely to the richness of the subject.

One visualizes a surface in 3-space as a curved sheet like the surface of a ball or a rolled-up piece of paper. One can describe them in terms of equations just like curves in the plane. On the surface of a sphere (depicted in Fig. 17), we have local coordinates near some point (for instance, one can think of latitude and longitude), which we think of as a flat two-dimensional plane with coordinates u and v superimposed on the curved surface, and we can study curves, paths of ships, airplanes, and so forth in terms of curves in the coordinate system. The (x,y,z) coordinates and the equation $x^2 + y^2 + z^2 = R^2$, where R is the radius of the sphere, are *extrinsic coordinates*, that is, they are outside (*ex*) the surface. The coordinates (u,v) on the surface are intrinsic (*in*), that is, on the surface itself.

Intrinsic geometry is the circle of geometric ideas and relationships that one can formulate and understand on the surface itself. *Extrinsic geometry* relates to the relation of the surface to the exterior three-dimensional world. For instance, if we considered the motion of a projectile from the surface of the sphere, like a rocket from the surface of the earth, then its path in space is *extrinsic* to the geometry of the surface. The projection of the curve onto the surface is an *intrinsic curve* that partially describes the extrinsic motion.

Riemann introduced in the 1850s the notion of a *manifold*. A manifold of two dimensions is a space (which we schematically draw just as

Figure 16. *Spherical geometry*

segment of great circle joins two points

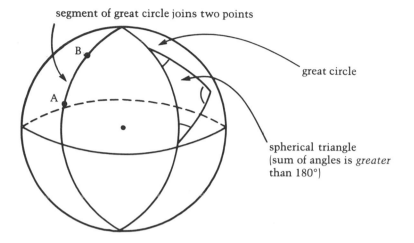

great circle

spherical triangle
(sum of angles is *greater* than 180°)

Figure 17. *Coordinates on a spherical surface*

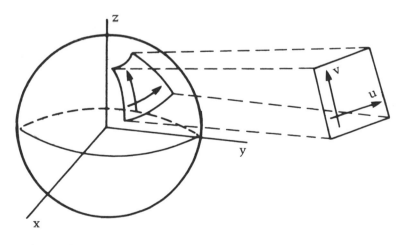

sphere of radius $R = \{x^2 + y^2 + z^2 = R^2\}$

we did above for the surface of a sphere) in which near every point we have local coordinates *(u,v)*, but *without* supposing the surface is specified in some ambient 3-space. This was an abstraction of familiar surfaces in 3-space. In fact, one can describe (as was shown some decades later) two-dimensional manifolds of this sort that one *cannot* realize in 3-space.

In this manner the set of geometric objects was considerably enriched. One can also envision manifolds of three, four, and higher dimensions where we specify that near each point there are three coordinates *(x,y,z)* for a three-dimensional manifold and four coordinates *(x,y,z,t)* for a four-dimensional manifold, and so on. These ideas are often difficult for non-mathematicians to grasp, and the notion of four-dimensional (and higher) manifolds or spaces took on a mystical air in the popular press in the late nineteenth century and on into the twentieth. For the mathematician these spaces, which were constructs of the imagination just as a symphony is for a composer, could be studied in exactly the same manner as one studied other parts of mathematics. Indeed, the geometry of the Greeks was equally abstract. It is just that one has everyday objects that one can visualize when one talks about a plane, for instance. A four-dimensional plane, on the other hand, is just as viable to the mathematician but has no counterpart in everyday life. On the same note, many concepts of modern science from quarks to semiconductors to lasers have no everyday counterpart.

Each system of thought, be it mathematical, physical, philosophical, or otherwise, builds upon the previous systems of thought as underpinnings, and often these systems become far removed from our physical childhood senses (for example, the notion of speaking on a telephone can be taken for granted only after one gets used to it; it certainly does not correspond to our "everyday experience" of talking directly with someone).

When Riemann invented the notion of a manifold, one was able to apply the notion of intrinsic geometry, since to study geometry intrinsically on a surface or on a higher-dimensional manifold, it is not necessary to know how or whether the manifold is embedded in some ambient Euclidean space. In the late nineteenth century these ideas were carried out by the Italian geometers Levi-Cevita and Beltrami, among others. They showed, starting with the initial ideas of Riemann, how one could study systematically geometry in an *n*-dimensional manifold for any dimension *n*, and this included questions of the following sort: how to measure distance on the manifold (like how to measure distance on the surface of the earth), how to measure angles, how to tell when a (small piece of a) line at one point is parallel to a similar "infinitesimal" line segment at another point, how to measure curvedness of the manifold, and so forth.

Are the two directions at the points *P* and *Q* parallel *on* the manifold? Levi-Cevita made sense out of this question. Thus, we see that the same type of geometric questions asked by the original Greeks were being formulated and answered in this higher-dimensional and more abstract setting. The stage was now set for the dramatic events of the twentieth century in the form of relativity theory and quantum theory.

Figure 18. *Extrinsic curve and its intrinsic projection onto the surface*

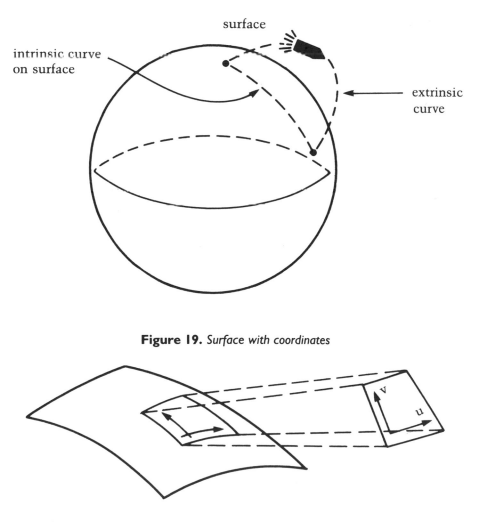

surface

intrinsic curve
on surface

extrinsic
curve

Figure 19. *Surface with coordinates*

Figure 20. *Parallelism*

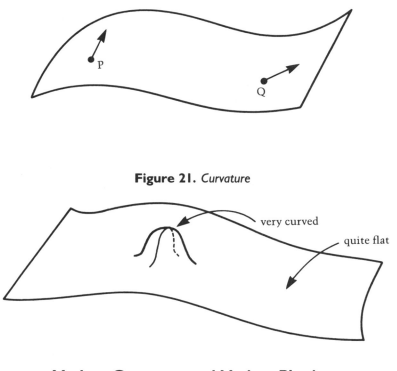

Figure 21. *Curvature*

Modern Geometry and Modern Physics: An Interwoven Mosaic

In the previous section I discussed the notions of manifold and intrinsic geometry as being two fundamental new creations of nineteenth-century thinking. These were not the only ones, and in the early part of the twentieth century there were more. I will summarize briefly the modern branches of geometry that have now evolved and their relation to physics. Before I do this, we need one key ingredient that goes back to the early nineteenth century: the discovery of *complex numbers* by Argand, Gauss, and others. The fundamental problem was understanding the solutions of the equation

$$x^2 = -1.$$

If we solved the equation $x^2 = 1$, we would write the solution as $x = \pm\sqrt{1}$, but for $x^2 = -1$, we would have to write something like $x = \pm\sqrt{-1}$. Since the square of any ordinary number is positive, what could one mean by the square root of a negative number? This was a dilemma of mathematicians for many centuries, and the traditional approach had been to say that such equations were impossible to solve, and one avoided the issue. But the inventive mathematicians of the early nineteenth century conceived of a geometric generalization of the usual number system in which one could make sense of this issue. They represented $\sqrt{-1}$ as being a *rotation* in the plane of 90°. We imagine the real number line as lying in the plane along the x-axis as pictured in Figure 22, and we let i denote rotation by 90° in the counter-clockwise direction. Now apply i to the number 1, and we get the point in Figure 22 labeled $i \cdot 1$ (with coordinates (0,1)). Perform the same rotation again, and we have $i \cdot i \cdot 1 = -1$. Thus, we see that the "product of the rotation by itself" (which we can denote by i^2), multiplied by the number 1 gives -1. Since we would want any number multiplied by the number 1 to be itself, we see that

$$i^2 \cdot 1 = i^2 = -1,$$

and thus we have found an object (the rotation) that performs as we would want $\sqrt{-1}$ to perform algebraically. It is a geometric object relating to the plane and requires going away from the number line contained inside the plane to be explicable. Today we consider complex numbers as being combinations of the usual numbers with the *imaginary unit i* of the following form:

$$z = x + yi,$$

where x and y are real numbers. We associate the complex number of the above form with the point (x,y) in the plane as depicted in Figure 23, where x and y are coordinates of points in the plane in the style of Descartes. Numbers of the form $x + 0 \cdot i \, (= x)$ are the usual real numbers, while numbers of the form $0 + y \cdot i$ are called *imaginary*. Consider a specific imaginary number, say $3i$, for instance. Then one sees that

$$(3i)^2 = (3)^2(i)^2 = (9)(-1) = -9,$$

and thus that imaginary numbers have the property that their square is negative, which is what was sought. Thus, we see that the notion of "numbers" was enlarged by extending to the complex numbers. Later Hamilton went one step further with his quaternions, which were like complex numbers but did not satisfy the commutative law, as was discussed in the previous section. (To describe a quaternion in terms of real numbers one needs four real numbers instead of the two required for

Figure 22. *Rotation by 90° as imaginary unit i*

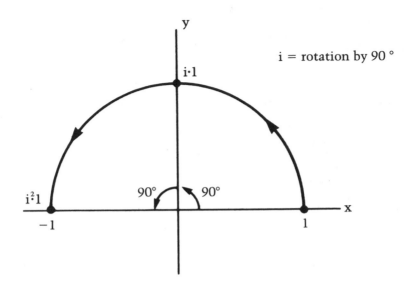

Figure 23. *Complex numbers*

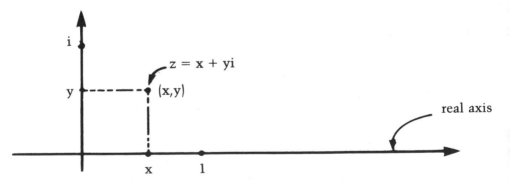

complex numbers). The development of complex numbers led to a branch of mathematics called "complex variable theory" or "complex function theory," which was one of the most important mathematical developments of the nineteenth century. This relates to our theme of geometry in that the complex numbers are describable in geometric terms as we saw above, but in addition one can consider geometric manifolds where the local coordinates are *complex numbers* in a suitable manner, and these are called *complex manifolds*. The study of complex manifolds has been of considerable interest in the mid-twentieth century and is still an active branch of research. Thus, we have an even more subtle interaction of geometry and algebra.

What are the major branches of geometry today that transcend the heritage of the Greeks at the beginning of the nineteenth century? There are four major branches that can be identified.

(1) *Intrinsic geometry on manifolds*. This is called differential geometry normally and was discussed extensively in the previous section. It includes the plane geometry of the Greeks as well as the original non-Euclidean geometries of the nineteenth century.

(2) *Infinite dimensional geometry*. Here one considers space with an infinite number of coordinates in a systematic manner, generalizing usual notions of line, plane, perpendicular, rotation, and so forth to this setting. An important result is the infinite-dimensional version of the principal axis theorem, due to Fermat in two dimensions. In the two-dimensional case, Fermat showed that if an ellipse is given in a plane with a coordinate system, then one can find a translation and a rotation of the plane so that the new coordinate axes are centered at the center of the ellipse and are passing along the major and minor axes of the ellipse as pictured in Figure 24. This was the problem solved by Fermat, which is taught to our high school students today. The infinite-dimensional version of it is called the *spectral theorem* and involved the work of various mathematicians in the early part of this century. In an infinite-dimensional space of a suitable type, one is given explicitly a generalization of an ellipsoid (some sort of quadratic type of function), and the problem is to determine the principal axis directions and the distortions in those directions. Then the directions and scalings (relative distortions) in those directions can be used as a suitable coordinate system for the original space. Such coordinates are often much more suitable for the problem at hand, just as in the case of a classical ellipse. (For instance, if we want a suitable coordinate system for the earth and sun, there are good coordinates that are adapted to the elliptical motion of the earth about the sun, and finding those coordinates is what we are talking about.)

(3) *Topology*. Topology grew out of the study of the geometry of manifolds, and in particular from the desire to understand quantitatively the global behavior of a manifold. The fundamental question asked in topology is: when can two manifolds be continuously equivalent, that is, when can one manifold be continuously deformed into another? In Figure 25 we see examples of three surfaces in 3-space that are not continuously deformable into one another. They have different numbers of

Figure 24. *Principal axis theorem of Fermat in 2 dimensions*

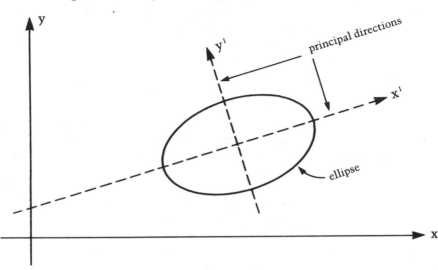

"holes" in the surfaces, and the "number of holes" turns out to be a precise measure of the different classes of manifolds, which are equivalent to one another in the sense of topology.

The topology of surfaces was developed in the latter part of the nineteenth century, starting with the work of Riemann and with a key contribution by the Italian mathematician Betti. At the beginning of the twentieth century, Poincaré in Paris formulated what became the major fundamental research problems in topology for the twentieth-century researchers. The questions of topology concern a much less rigid geometry than that of the original rigid motions of the Greeks. For instance, in the plane a circle, a triangle, and a rectangle would all be equivalent from a topological point of view. Philosophically, one proceeded in a similar way as the Greek geometers did; just the nature of the geometric objects and how they should be considered equivalent had changed. In topology one has a general class of geometric objects called n-dimensional manifolds, and one wants to classify them into continuously equivalent classes, much as triangles in the plane are classified by being either congruent (equivalent after a rigid motion) or similar (same rigid motion, but allowing a change of scale). There is also a hierarchy of classifications, just as there is in biology with genus, species, and so on. The actual problem of classification is quite subtle. A major advance in the classification of 3-manifolds was achieved in the past decade by W. Thurston of Princeton University, along with others who made impor-

Figure 25. *Topologically inequivalent surfaces*

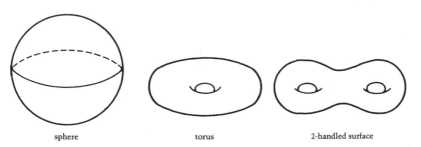

sphere   torus   2-handled surface

tant contributions. Much more recently the problem of classifying 4-manifolds was completely and spectacularly resolved in 1982 by M. Freedman of the University of California at San Diego, and this solved one of the major problems posed by Poincaré at the beginning of this century.

(4) *Singularities.* We have talked at length about manifolds, but sometimes we have an object like a cone pictured in Figure 26. In this case it is impossible to put a coordinate system in any sensible fashion at the apex of the cone. In general, we would call such a point for which we could not find good local coordinates a *singular point* of the space.

The study of singularities is an important part of modern geometry and is studied from various points of view by different mathematical disciplines: algebra, complex coordinate systems, topology, and differential equations, to mention a few. But I will not attempt to say any more about it in this essay, except that it is a very vigorous part of modern mathematical research.

All of these four types of modern geometry play an important role in contemporary physical models of the universe. For instance, Einstein's theory of general relativity is conceived of as a four-dimensional manifold called space–time. Its local coordinates are three spatial coordinates and one time coordinate (x,y,z,t). On space–time there is a *distance function* prescribed with the property that the *shortest distance between points* corresponds to the paths of moving objects (for example, people, planets, stars, spaceships, galaxies, and so on). These paths are called geodesics, and this name harkens back to the first study of differential geometry by Gauss, which was related to his studies of *geodesy*, the theory of measurements of the surface of the earth. The geodesics of the surface of the earth are great circles, but the geodesics of Einstein's space–time are the paths of *all* moving objects, a much more universal concept. So understanding completely the geometry of space–time

Figure 26. *Singular point of a cone*

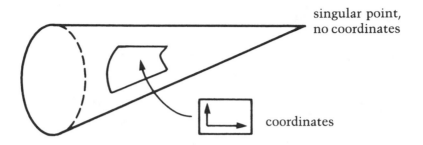

singular point,
no coordinates

coordinates

would certainly give us a lot of information about the world around us! This is a rather impossible task for a given generation of human beings, but it is nevertheless a very useful model for asking more limited questions about the universe around us.

The *curvature* of Einstein's space–time represents the amount of gravitational energy (or attractive force) in the space–time near a particular point of curved space–time. Thus, in Einstein's view the portion of space–time occupied by the sun at a particular point of time would be more curved than, say, near the planet Pluto. It was the bending of light rays of a distant star as they passed close to the sun (and hence to the sun's gravitational attraction) that was predicted by Einstein in 1916 and verified by Eddington in 1919, which was the first major justification of his point of view. The Newtonian picture of gravitation had no interaction of light rays with gravity at all. Thus, one has a geometric model of the universe, but quite different from the epicycles of Ptolemy (who, as we recall, was trying to understand the motion of planets, a far cry from the behavior of a stream of photons in a light ray from a distant star). All of the ideas of Gauss (intrinsic geometry), Riemann (manifold, curvature), and Levi-Cevita (parallelism) go into Einstein's model of gravitation, and it is currently the most widely accepted theory of gravitation available. There is certainly the possibility that a later scientist will be able to find an improved version that will not only contain both the Einsteinian and Newtonian theories but will account also for phenomena from quantum theory (which is not yet the case). We now turn to this second fundamental discovery of twentieth-century science.

In quantum mechanics one uses complex numbers, the geometry of infinite-dimensional spaces, and the spectral theorem as important tools in understanding the world of the atom and the nucleus. In the small world of the nucleus of an atom, radiation occurs in clumps of discrete units (quanta). The light rays of everyday life are composed of enormous numbers of these small quanta so that it seems to come in a continuous flow. This is analogous to the seeming continuous motion of a movie, which is, in fact, a sequence of distinct photographs. Only if we look on a small enough scale (analogous to slowing the film down), can we perceive this discrete nature of radiation and light. The discrete levels of radiation of the element hydrogen when it is burned are called the spectral lines, and these appear discretely in a laboratory or can be perceived in the light of distant stars. This spectrum of hydrogen, which is observable in an experimental sense, corresponds in a precise manner to the infinite set of principal distortions on a suitable "ellipsoid" in an infinite-dimensional space. Such descriptions of physical phenomena of this nature are part of what is called today *quantum mechanics*. It is a very accurate and quite useful description of the world of the atom.

The notion of topology plays an important role in both quantum theory and relativity theory. We still do not understand what the global nature of our Einsteinian space–time universe looks like from a topological point of view, but there are various important speculations in this matter. For instance, one important question is whether the universe is "open" or "closed." In ancient times the analogous question was whether the earth was flat ("open") or spherical ("closed"). These two types of universes would be distinct topologically as well as in many other ways, and contemporary research in astrophysics and cosmology is attempting to give an answer to this question. In quantum theory the simple distinction between discrete and continuous is topological in nature. In recent models of elementary particles, one has associated various integers to elementary particles called "quantum numbers," and these are nothing but topological "invariants" in certain geometric models of these particles. A topological invariant is a quantity (often an integer) that is associated to a geometric object and that is invariant (unchanging) under continuous deformations. For surfaces in 3-space the number of holes in the surface (as illustrated in Fig. 25) is an example of a topological invariant. In the physical context one has the topological invariant having a physical interpretation.

The notion of singularity comes in when one considers the space–time universe of Einstein as a four-dimensional manifold but where there is a definite notion of time. The big-bang theory asserts that there was an *initial time* when the universe was essentially an extremely dense point mass, which expanded rapidly to its present size. This initial point is a singular point of the space–time universe, as there is no coordinate system valid on all sides of this point, much as the apex of a cone is a singular point for a cone; this is depicted in Figure 27.

One of the major problems of theoretical physics today is to understand the relation between quantum theory (the physics of the very small) and relativity theory (the physics of the very large). Both of these

Figure 27. *The evolution of the universe*

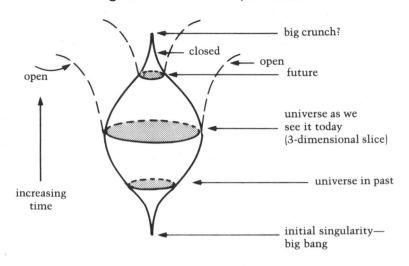

theories have as limiting cases the classical physics of Newton (the physics of everyday sizes), but normally one must assume either one model or the other (or the limiting classical case if that suffices). What is missing is some unified theory that will contain all of these theories as special cases and that will allow one to go back and forth without having to make ad hoc assumptions. There are some limiting situations in which it will be necessary to understand both theories at the same time; for instance in the vicinity of the initial singularity of the big bang, one has the situation that the gravitational field is very intense (hence the need for Einstein's theory) and things are quite small (hence the need for quantum theory). Similarly, in the neighborhood of a black hole, one is forced to consider both points of view at the same time. A black hole is an astronomical object that has such a strong gravitational field that the light rays return to the matter emanating them and are not visible to an outside observer (thus the "black" appearance).

Modern scientists have differing notions as to how this unification might proceed. For instance, Hawking has developed a notion of "space–time foam," where the continuum space–time would appear to look like a sea of discrete "bubbles" if viewed though a sufficiently high-powered "microscope" (down to the order of 10^{-33} cm in width, in contrast to the size of a proton of about 10^{-13} cm in width!). The bubbles are in fact very tiny four-dimensional manifolds that are complex manifolds of specified types and with specific properties.

Penrose introduced in 1967 his twistor theory as a new type of background space in which to view the universe. This replaces the four-

Figure 28. *Space-time foam of Hawking*

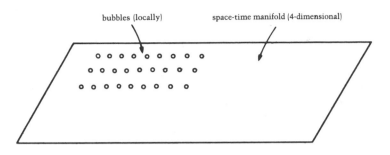

dimensional manifold of Einstein by a 3-complex-dimensional manifold in which the physical world can be modeled. Twistor theory has been shown to have deep links both with quantum theory and with relativity theory and may represent a type of mathematical structure in which both theories could be understood simultaneously.

A third program of various physicists is that of supergravity, a formal generalization of the algebraic aspects of Einstein's theory, which incorporates features found in quantum theory. The geometric foundations of this type of theory are only very recently beginning to be understood, and there is considerable activity in this direction both from the point of view of mathematicians (there is some new geometry to be understood) and from the point of view of physicists (how does the proposed physical model really work?).

At present none of these theories gives any real solution to the dilemma, and one of the principal reasons is the fact that there is no experimental evidence available to distinguish one theory from another at this point. Experimental physicists have carried out many and varied experiments showing that Einstein's theory and the quantum theory are valid over a broad range of their applicabilities, and both are capable of giving accurate predictions of physical behavior in their appropriate ranges. However, there are at present no experiments that have been conceived with contemporary tools that will measure adequately at the overlap regions of validity of both of these theories. Thus, we have no way of testing any proposed theory of quantum gravity (as such a unified theory is generically called today). We seem to be in a holding pattern on these questions because these problems have been in the air since 1925, when quantum theory was ushered in and relativity theory was in its infancy, having been brought forth in 1916, as mentioned earlier.

This brings us to the end of an excursion through the historically evolving geometric theories of the mathematicians and the applications of these ideas by scientists attempting to understand an increasingly complicated world. No doubt, some of the abstract ideas of the current generation of geometers will play some role in the twenty-first or twenty-second century's view of the universe. Hindsight tells us that it will be quite unlikely if anyone could *predict* which of the currently active areas of geometry will play an important role. On the other hand it seems very likely that some of the current geometric theories being developed by mathematicians today will play a significant role in the future understanding of the universe.

The Origins and Development of Medieval Mythography

From Homer to Dante

Jane Chance

When Dante wished to explain to his patron Can Grande the rationale for his great work, the *Commedia*, he pointed to the multiple allegorical levels operating in the following Psalm text (the Douay translation, 113:1–2; 114:1–2 in the King James) based on the story in Exodus: "When Israel came forth from Egypt, the house of Jacob from a people of alien tongue, Juda became his sanctuary, Israel his domain." The Psalm can be interpreted as follows if the literal level is distinguished from the allegorical levels:

> Now if we look at the letter alone, what is signified to us is the departure of the sons of Israel from Egypt during the time of Moses; if at the allegory, what is signified to us is our redemption through Christ; if at the moral sense, what is signified to us is the conversion of the soul from the sorrow and misery of sin to the state of grace; if at the anagogical, what is signified to us is the departure of the sanctified soul from bondage to the corruption of this world into the freedom of eternal glory.[1]

The three allegorical levels discussed by Dante are, in addition to the historical or literal level, the allegorical, moral (or tropological), and anagogical. These last three levels as a group are distinguished from the literal level because, explains Dante in the Letter to Can Grande, "allegory" derives from the Greek *alleon*, which in Latin is *alienus*, "belonging to another," or *diversus*, "different."

Dante's treatment of this biblical text may seem bizarre and unusual, even peculiarly idiosyncratic, to a modern reader, but in the Middle Ages such treatment was the product of thousands of years of exegesis (explanation or annotation, often interpretation) of text derived from Greek and Hebrew models and would have been immediately comprehensible by other poets and scholars.

As a method of interpreting sacred texts, such exegesis derived originally from ancient Hebrew explication of the Bible, adopted by the early Greek Fathers of the Church to justify the existence of persons or events in the Old Testament by understanding them as types— anticipations or foreshadowings—of persons or events narrated in the New Testament. This rationalization was perceived as necessary because early Christianity denigrated Hebraism as the religion of the unconverted, fallen heathen; yet, if the Old Testament revered by the Jews was written by God, should not the Old Testament also be studied by Christians? It might be if Moses, for example, is understood as a type of Christ in the strength of his faith.

A different method of interpreting sacred texts from which later medieval exegesis also derived more immediately and specifically was propounded by the Church Fathers. In an attempt to justify apparently immoral or blasphemous passages, particularly in the New Testament, the Fathers explained that a text might have another, more spiritual meaning. St. Augustine might cite as an example of double levels the anointing of Christ's feet with precious and sweet-smelling nard, which literally might seem sensuous and indulgent, but which figuratively means the pleasant odor of fame from the good works performed in a good life. This method came to be used to interpret secular literary texts, such as Dante's *Commedia*, by poets or scholars who saw their poems as having a religious context. For the Middle Ages the similarity between secular and sacred texts derived from the texts' common use of multiple levels of meaning. Thus, Dante wished Can Grande to understand that his *Commedia* on the literal level concerns the state of souls after death, whether in Hell, Purgatory, or Paradise, but on the other three levels concerns the consequences of man's free will, either the rewards or punishments of justice. The difference between the *Commedia* and the Bible, for exegetical purposes, is great: the former text treats of pagan figures and monsters, especially in the *Inferno*, not found in the latter. For the exegesis of mythological heroes and gods Dante needed to turn to a very different allegorical tradition from that of the ancient Hebrews and Fathers of the Church. Such exegesis was derived from ancient

Greek rationalization of the epics of Homer and is termed "mythography," or the moralization and allegorization of classical mythology.

Mythography differs from mythology chiefly in its form: mythology is a unified system of myth, often in narrative form, whereas mythography is an explanation and rationalization of one or more myths, often in didactic form. Indeed, for Homer, the earliest known user of myth in the Graeco-Roman world, *mythos* signified either speech, or unspoken words and thoughts as contrasted with deeds,[2] although the word "mythos" could also signify true or false story, rumor.[3] A definition possibly clearer to the modern reader is the later Aristotelian definition of *mythos* as "plot," or, in Northrop Frye's terms, the "verbal imitation of ritual."[4] Mythography, in contrast, is the interpretation of myth—its meaning, or what the ancient philosophers termed its *hyponoia,* or "undermeaning." As exegesis, it employs the tool of allegorical interpretation, which belongs more to the ancient exegete, whether philosopher or grammarian, than to the poet, although Roger Hinks has noted that

> Allegory stands, as it were, midway between poetry and prose: in its creative aspect it is the poetic rendering of a prosaic idea; in its interpretative aspect it is the prosaic rendering of a poetic image. Like the daemonic faculty in the words of Diotima, it reveals the ways of man to the gods and the ways of the gods to man.[5]

Mythography in the Middle Ages also depended on allegorical fabrication in a manner remarkably similar to the ancient technique, but elaborated and refined. The resurgence of interest in the classics in the twelfth-century renaissance was accompanied by a renewed use of mythography to explain classical mythology. At that time the term *allegoria* assumed a more specific application: it was used to refer to scriptural story containing the Word of God, whereas the terms *involucrum* or *integumentum* ("cover," "cloak") were used to refer to classical myth containing philosophical ideas about the World Soul and other kinds of moral truth.[6] Hence, instead of "allegorical interpretation" we should substitute "integumental interpretation" for later mythographic exegesis to distinguish the two terms, although for the purposes of this study they will be treated equivocally. The concept of *integumentum* was created by twelfth-century scholars to discuss classical fable, in particular from a Neoplatonic moral interpretation of the Bible derived from St. Augustine. He had suggested in *De doctrina Christiana* that the student must not read a text carnally, that is, by the letter only, but instead by the spirit, that is, according to its figurative meaning. This approach implies that a text, particularly a biblical text, like man has a body and soul. Such an idea was made explicit by Bernard Silvester in the twelfth century when he declared that for myth the integument cloaks truth just as the body does the soul and hence should be understood by means of the intellect. He defines *integumentum* as a "type of exposition which wraps the apprehension of truth in a fictional narrative, and thus it is also called an *involucrum*, a cover."[7] John of Salisbury in the same century described *involucrum* as the fictitious cover concealing divine prudence and, like Bernard, also cited the *Aeneid* as an example.[8] Further, scholastics of the twelfth century multiplied these definitions endlessly: the letter (*littera*) is the surface meaning of a poem, also known as history (*historia*)—defined in a very specialized sense as "the grammatical construction of the text"; a synonym for this "outer covering" of the poem is the *cortex* (also known as *sensus, pallium, tectorium, tegmen, paille* in Old French, *chaf* in Middle English), as distinguished from the doctrinal or moral theme or thesis, the truth hidden under the *cortex*, which was also called the *nucleus* or *sententia, fruyt* in Middle English.[9] The use of these terms and these tools to probe beneath the literal meaning of a narrative was part of the grammarian's task in understanding (in particular) a Latin classic; he often turned to a mythographer, who explained or rationalized morally classical myths in order to understand the hidden meanings.

Mythography was first used to defend apparently blasphemous passages in Homer, especially concerning the battle of the gods in *Iliad* 20.67, but, remarkably, it endured and flourished through the Middle Ages, in Dante and later poets, and into the sixteenth and seventeenth centuries in the visual arts and letters. Why it endured as a method is a question that goes hand-in-hand with another, which is, why pagan gods and goddesses and other apparently blasphemous materials continued to appear in medieval poetry.

This essay will define this method of allegorizing classical myths. First, the development of the method will be traced from ancient Greek practice in Homer to the early fourteenth century, when Dante wrote the *Commedia*; second, a historical explanation of why the method developed in this way will be offered; and third, a late medieval poet, in this case Dante, will be examined to reveal how he used the techniques in writing his own epic and how he drew upon earlier mythographic sources to do so.

I

In this first part of the essay, I shall explain how the mythographic method developed from ancient Greek practice in Homer to the early fourteenth century, when Dante wrote his Christian epic. Its evolution was kept alive throughout these many centuries for basically one reason: the supporters of poets—essentially philosophers and grammarians—wished to defend their use of material that was apparently blasphemous, by pagan or Christian standards. That is, if Homer depicted Greek gods engaged in immoral or licentious activities, or if Dante portrayed immoral pagan figures occupying or ruling the circles and *bolge* (pockets) of Hell, there must have been some underlying, hidden meaning to justify these poets' depictions. Homer and Dante supply appropriate examples for our survey because both demonstrate such immoral activity of the gods in their depictions of Hades, or the Underworld. But they serve as useful examples for another, chronological reason: Homer's poetry was responsible for initiating the practice of mythography in Greek antiquity, and Dante's poetry was responsible not so much for ending it—it continued for several centuries—as for using specifically medieval versions of mythography in a unified way that no later poet would.

In ancient Greek justifications of Homer there were basically three kinds of hidden meaning that licentious material could conceal—the natural or physical, the moral, and the grammatical.[10] The physical undermeaning, or *hyponoia*, referred to natural forces or phenomena, the moral *hyponoia* referred to human faculties or qualities, and the grammatical *hyponoia* referred to the philosophical reality of a name. In the Greek period the physical meaning was initially the most prevalent, followed by the other two, while in the Middle Ages all three coexisted from the earliest examples. In addition, in the Middle Ages the grammatical and moral senses came to be conflated in the period up to and including the Carolingian period (eighth and ninth centuries) and supplemented (even supplanted) by a fourth type of meaning, the Neoplatonic and Christian, between the twelfth and fourteenth centuries.

The origin of these types stemmed from the defense of Homer and Hesiod. The eighth-century Homer, author of the epic poems *Iliad* and *Odyssey*, and Hesiod, author of a short epic treatise, *Works and Days*, were both criticized by early philosophers for their descriptions of the gods engaging in immoral behavior. Such criticism resulted in the fantasy that both poets received appropriate punishments for their infractions in Hades: Pythagoras of Samos (582–500 B.C., the dates set by Diogenes Laertius), who supposedly descended to Hades, claimed to have witnessed "the soul of Hesiod bound fast to a brazen pillar and gibbering, and the soul of Homer hung on a tree with serpents writhing about it, this being their punishment for what they had said about the gods. . . ."[11] Other criticisms seem equally harsh. Xenophanes sternly declared that both "Homer and Hesiod have attributed to the gods all things that are shameful and a reproach among mankind: theft, adultery, and mutual deception."[12] Heraclitus (540–470 B.C.) lambasted Homer

("Homer deserves to be flung out of the contests and given a beating")[13] and Hesiod to a lesser extent ("Hesiod is the teacher of very many, he who did not understand day and night: for they are one").[14]

In the century and a half after Homer's death, pre-Socratic philosophers first offered specifically physical allegorical interpretations in defense of the eighth-century B.C. Homer (and Hesiod). One of the earliest defenses occurs in support of 1.590 and 15.18 of the *Iliad*. In the first instance, Hephaestus (Vulcan, god of fire in Roman mythology) says to Hera that Zeus has hurled him below the divine threshold, and in the second, Zeus elaborates on (and justifies) this passage by explaining to Hera that this means that Hephaestus had been sent to earth because the gods could not free him from being bound with two golden chains to Hera, who hung from the aether and clouds on high. Both passages are interpreted by early defenders of Homer as cosmological descriptions. For example, Pherecydes of Syros in the seventh century or mid-sixth century B.C. in *Heptamychos* (*The Seven-Chambered Cosmos*) explains that an underworld for the punishment of the gods exists beneath our known world: "Below this part of the world is the Tartarean part; its guardians are the daughters of Boreas, the Harpies and the Storm-wind. Thither does Zeus banish any god who commits an act of lawlessness."[15] According to Origen's recitation of this myth in *Contra Celsum*, pride or arrogance is the specific act of lawlessness that results in this punishment. Later, Celsus would agree with this cosmological interpretation: he understood that Zeus is really God and Hera is Matter, that the earth beneath the "divine threshold" is in reality the Underworld, and that Hephaestus was sent there as punishment for his arrogance.[16]

Another passage criticized by early philosophers involved *Iliad* 20.67, on the partisan involvement of the gods in the Trojan War. Theagenes of Rhegium (ca. 525 B.C.), the first defender of Homer to use allegory explicitly, provided two explanations, natural and moral, that would develop into separate approaches later in the evolution of mythography. In a *scholium* on *Iliad* 20.67 cited by Porphyry in the *Theomachia*, he explains the battling of the gods as representative of physical conflict among the elements: just as there exists a natural conflict between contrary elements, hot versus cold, light versus heavy, so also water quenches fire, with fire expressed by Apollo, Helius, and Hephaestus, and water by Poseidon and Scamander (and the moon by Artemis and the air by Hera).[17] He also suggests such battling among the gods can be explained by mental (psychological) oppositions. Athena is Fronesis (Wisdom), who wars with Ares (Foolishness); Aphrodite is Desire, who wars with Hermes (Logos).

The physical interpretations of Homer and Greek myths in general became better developed in the latter half of the fifth century B.C., especially in the hands of the Ionian, Sophist, and Cynic philosophers. Several Ionians argued that Homer's purpose had been to propound virtue and justice (and, therefore, to justify whatever apparently immoral divine activities were portrayed).[18] Anaxagoras of Clazomenae (ca. 460 B.C.) viewed Zeus as mind and Athena as art or technical skill,[19] and the rays of the sun as the arrows of Apollo.[20] He says, "We give the name Iris to the reflection of the sun on the clouds. It is therefore the sign of a storm, for the water which flows round the cloud produces wind or forces out rain."[21] His pupil Metrodorus of Lampsacus (d. 464 B.C.; none of his writings are extant) claimed that neither the gods nor the heroes existed but were introduced by the poet for artistic reasons, "referring all to physiology" (*omnia ad physiologiam revocans*) so that Hera, Athena, and Zeus could be equated with "the parts of nature and dispositions of the elements" (*naturae partes et elementorum dispositiones*).[22] Similarly, the heroes of the *Iliad* are identified with physical phenomena— Agamemnon with aether, Achilles with the sun, Helen with the earth, and Paris with the air—while the gods are identified with parts of the human body—Demeter with the liver, Dionysus with the spleen, and Apollo with the gall.[23]

The Sophists stressed etymological interpretation of myth even as they explained the gods as aspects of the natural world. Democritus (ca. 420 B.C.) equated Zeus with air,[24] and noted that "Tritogeneia," literally thrice-born, means that wisdom, or Athena, consists of three parts.[25] Prodicus of Ceos, in the latter half of the fifth century, also used etymology to explain how myths cloak truth: he declared that the ancients regarded the sun, moon, rivers, springs, and all other things beneficial to man as gods because they served man, as was the case with the Nile for the Egyptians. Thus, bread was understood as Demeter, wine as Dionysus, water as Poseiden, fire as Hephaestos.[26]

That such explanations were accepted and believed is clear from the philosophical satires of Cynic philosophers like Epicharmus of Syracuse (ca. 550–460 B.C.) in his comedies and memoirs: Jupiter is said to be

> Air; who is wind and clouds, and afterwards rain, and from rain comes cold, and after that, wind and again air. Therefore these elements of which I tell you are Jupiter, because with them he helps all mortals, cities and animals.[27]

In the sixth century Plato advanced etymological allegorism as a reflection of both moral and physical allegorism. In the *Cratylus* he begins with Homer, as so many of the philosophers do. Plato argues that Homer correctly attributed names to particular things, as witnessed in the denotations of such names, for example, as "Xanthus" for the name of a river rather than "Scamander" (391D–E), and the connotations of such names, for example, of Agis as "leader," Polemarchus as "warlord," and Acesimbrotus as "healer of mortals" (394C).[28] But Plato also argues that names reveal a moral reality in addition to etymological connotations. Thus the name "Agamemnon" ("admirable for remaining") mirrors his character as "one who would resolve to toil to the end and to endure, putting the finish upon his resolution by virtue. And proof of this is his long retention of the host at Troy and his endurance" (395A). The discussion concludes, after additional examples are cited, with a display of how etymological allegorism can also support a physical interpretation of the gods: the names of the gods are representative (for the earliest Greeks, at least) of the sun, moon, earth, stars, and sky because the Greek word for "god" (θεόσ) comes from the fact of their constant running (θεῖν).

Such etymological and physical allegorism culminated in the practices of the Stoics, who in the fourth to third centuries B.C. created a system of allegorical details in support of Homer: they assumed that Homer wrote with an understanding of Stoic physical and moral dogma.[29] Zeno of Citium (340–265 B.C.), father of the Stoics, rationalized Homer's use of the gods by showing how all fit into an orderly natural schema wherein names signify natural forces.[30] The major gods represent the regions of the universe: Juno is air, Jupiter, the heavens, Neptune, the sea, Vulcan, fire;[31] aether is a god and a principle of reason;[32] the Titans are the elements of the universe, as determined by the etymology of their name.[33] Zeno's teachings were expanded and developed by his followers, his pupil Cleanthes of Assos (b. ca. 300 B.C.) and Cleanthes' pupil Chrysippus (ca. 280 B.C.), as well as others.[34]

In Roman Stoic philosophy the process of mythography was continued by Cicero in his imaginary dialogue *De natura deorum*, an encyclopedia of Greek philosophy intended for Roman readers.[35] The Stoic defense of the gods in the second book is derived from the writings of Zeno, Cleanthes, and Chrysippus on the gods as natural forces, *physica ratio*. The immoral myths ingeniously conceal a unified cosmological theory indicating Stoic principles—for example, the fable of the castration of Caelus by Saturn signifies that the highest heavenly aether, that seed-fire that generates all things, did not require the equivalent of human genitals to proceed in its generative work (2.64). But Cicero depends heavily on etymological allegorism to make his points. When Jupiter puts Saturn in chains in an attempt to restrain his course and bind him in the stars' network, Cicero understands that *iuvans pater* ("helping father," our father and the father of the gods) is attempting to bind and restrain time (in that "Saturn" comes from "sated with years," *quod saturaretur annis*, 2.64). Similarly Stoic in their desire to embrace all natural and moral knowledge through the principles of etymological allegorism are the *Allegoriae Homericae* of pseudo-Heraclitus (of Pontus, at the time of Augustus) and the *Compendium theologiae graecae* of Cornutus (at the time of Nero).[36]

At this stage, the history of mythography splits. First, in the early centuries A.D. the Greek Alexandrine School and the early Church Fathers unwittingly preserved many of the Greek and Roman allegorical interpretations in their attacks on the pagan gods, especially in the writings of Tertullian (145–220 A.D.), Minucius Felix (ca. 210 A.D.), Arnobius (fl. 300 A.D.), Lactantius (ca. 306 A.D.), and most influentially, St. Augustine (354–430 A.D.). However, because many of their writings appeared in Greek, they had little direct effect on mythography in the Middle Ages.[37] Second, the various religious controversies of this period—between Christians and Jews and between Christians and pagans[38]—allowed apologists to adapt Stoic methods for rationalization of classical myths to Judaism or Christianity, thus preserving the methods of mythography for the Middle Ages. For example, Justin, in his *Apologia pro Christianis*, viewed Hercules, Bacchus, Bellerophon and Perseus all as types of Christ.[39] Theophilus saw the ancient myths of the Greeks as corresponding to biblical accounts, as in the floods of Deucalion and Clymenus that parallel the deluge of Noah.[40]

In contrast to the Christian denigration of classical myth and the application of Greek methods of allegoresis to scriptural materials in the period from the second to the fourth centuries, rationalization of classical mythology through mythography was chiefly continued by the Neoplatonist Macrobius in the late fourth century and various Neoplatonists in the twelfth century. Macrobius combines Stoic ideas drawn from Cicero's *Somnium Scipionis* (modeled on the closing episode in the tenth book of Plato's *Republic*) and Platonic ideas from the *Republic* in his commentary on the *Somnium Scipionis*. Herein he argues that the poet uses a fictitious style to treat truth in the *narratio fabulosa*, the fabulous narrative.[41] This fable, which advocates moral excellence, is used by philosophers to discuss the soul, the spirits of lower and upper air, and the gods more generally, at least as exemplified in the stories of Hesiod and Orpheus.[42] He denigrates the use of immoral stories involving the gods (adultery, the castration of Caelus by Saturn) or monstrosities as unfit for fabulous narrative. However, in the twelfth-century commentary on Macrobius by the Neoplatonist William of Conches, even this practice is rationalized according to the Stoic and Platonic principles of physical and moral allegorism, tinged with Greek Neoplatonism as filtered through medieval Latin commentators. (William has actually been credited with knowing Cicero's *De natura deorum* firsthand.[43]) Of the castration of Saturn by Jupiter (Macrobius is mistaken to refer to the castration of Caelus by Saturn, says William), he explains that the testicles signify the fruits of the earth ripened by the warmth of the upper element, or Jupiter, and that when cast into the sea (or figuratively into the human belly) give birth to Venus (or sensual delight).[44] The adulteries of Jupiter—not actually discussed by Macrobius—are revealed by William as nature myths: Jupiter (aether) sleeps with Semele (earth) in winter to give birth to Bacchus (the grape vines) twice, first by Semele (the vines becoming green when first impregnated by the sun's heat) and then by Jupiter's thigh (the vines putting forth grapes in summer).[45]

The moral, physical, and etymological allegorism of the ancients was thus reproduced by medieval commentators, and such mythography survived into the seventeenth century. One last type of allegorism was added and elaborated in the fourteenth century in particular: mythographers began to provide an ecclesiastical or Christian allegorism very like that of the early Greek Alexandrine School, which saw aspects of Old Testament events or the New—in Christ's nature or his life—in classical mythology. The thirteenth-century French voluminous *Ovide moralisé* detailed such correspondences and was summarized in part by Pierre Bersuire in his much more influential fourteenth-century *Ovidius moralizatus*. For example, Jupiter as the son of Saturn to whom the rule of heaven came by lot signifies God, the lord and master of heaven itself; Pierre cites Isaias 66:1, "Heaven is my throne," to tie the religious explanation firmly to the classical myth.[46] And yet he also supplies all of the other types of allegorism in his mythographic tract—often for the same examples provided earlier by other mythographers.

By Dante's time, in the early fourteenth century, the mythographic methods were well established. The idea of four levels of allegory—if not the same four levels—Dante derived from St. Jerome and St. Augustine, Cassian, Aldhelm, Hrabanus Maurus, Bede, John of Salisbury, and St. Thomas Aquinas.[47] Generally, those four levels depend on a distinction between the literal and the figurative (or allegorical) meanings. The concept of refining levels of meaning can be compared to the techniques advanced by the pre-Socratics, Platonists, and Stoic philosophers: the moral, physical, and etymological types of allegorism at least vaguely resemble respectively Dante's moral or tropological, anagogical, and allegorical levels. The physical meanings of the Stoics parallel the anagogical in that both reveal truths about the cosmos; the difference is that the Stoics were interested in Hades and earth, whereas the Christians, like Dante, are interested in the Underworld and Paradise. Further, the allegorical level, which reveals truths about Christ or Christ's life, can be seen in some way as similar to the etymological, which probes meaning through the origins and developments of words—suggesting an absolute moral reality outside of man. And, of course, the moral level of meaning, which sees the gods as expressions of human faculties or vices and virtues, remains the same whether Greek or Christian. While the preservation of ancient mythographic techniques in the Middle Ages may seem strange or surprising, nevertheless there were logical—and practical—explanations for such preservation.

II

The continuation and elaboration of mythography in the Middle Ages, to a large extent, was dependent on the preservation of the classics, which were used frequently in the schools. A medieval Christian student of Latin grammar often needed assistance in deciphering references to pagan gods and goddesses; such help was provided by lecturers who had studied glosses (often interlinear or marginal annotations of the text) or commentaries (often longer expositions that could stand alone and that introduce lines from the text for elaborate explanation) on the most important classics. Further, just as Greek mythography began with the inception of the mythological epic, particularly Homeric epic, so it continued in some way to be associated with that genre in the late antique era, even though it also came to be applied to other literary forms. Of the classical writers to whom medieval commentators were especially attracted—Terence, Virgil, Seneca, Ovid, Lucan, Statius, Augustine, and then in the period of late antiquity, Macrobius, Chalcidius, Martianus Capella, Boethius, and Fulgentius—the most important in parenting commentary traditions with mythographic significance were Virgil, Ovid, Lucan, Statius, Martianus Capella, Boethius, and Fulgentius. In addition, two other works with commentary traditions were modeled on the classical eclogues and epic of Virgil—the ninth-century *Ecloga Theoduli* and the fourteenth-century *Commedia* of Dante. This last work did not so much influence the Middle Ages as reflect in the many commentaries written on it the late medieval interest in the commentary tradition.

Of the classical works mentioned above as receiving special attention from medieval commentators, many are epic in form, a genre that itself derived from the Greek epics of Homer, or else were thought of as epic by medieval commentators and poets. Virgil's *Aeneid* demonstrates Homeric simplicity, unity, and consistency, whereas Lucan's *Pharsalia* and Statius' *Thebaid* were criticized by their contemporaries as faulty.[48] Ovid's *Metamorphoses* has only recently been identified as an epic,[49] and the later works—those of Martianus, Boethius, and Dante—can be viewed as epic only in a more medieval and hence allegorical sense. However, that all these Latin works were widely influential in the Middle

Ages, both in themselves and as source books for mythographic commentaries, is clear from the attention paid to them by late poets like Dante and Chaucer. Dante not only chose Virgil as a guide in the *Commedia*, but he also included Statius as a character in the *Purgatorio* (cantos 21 and 22) and Lucan, Homer, Horace, and Ovid (along with Virgil) as the five best classical poets in the *Inferno* (canto 3). Chaucer in *The House of Fame* (lines 1456–1512) positions various classical poets on pillars, including the epic poets Statius, Homer, Virgil, Ovid, Lucan, and Claudian, as well as writers on the Troy legend such as Dares, Dictys, Lollius, Guido delle Colonne, and Geoffrey Monmouth. That Chaucer regarded the legend of Troy as epic material in the tradition of the great classical epics is perhaps also evident in his request that his "litel bok," *Troilus and Criseyde*, "kis the steppes, where as thow seest pace/ Virgile, Ovide, Omer, Lucan, and Stace."[50]

The other three important late classical and late medieval works, Martianus' *De nuptiis Philologiae et Mercurii*, Boethius' *Consolatio Philosophiae*, and Dante's *Commedia*, transform elements of classical epic into uniquely medieval form. Martianus' *prosimetrum*, or mixture of verse and prose, which originated in the Menippean satire,[51] uses as its predominant verse form the dactylic meter of epic, invokes at the beginning the Muse of epic poetry, Calliope,[52] and borrows most heavily from the epics of Virgil, Lucan, and Ovid.[53] Boethius' *Consolatio* has been regarded as an imitation of Martianus' work, inferior only to Virgil in poetry and to Cicero in prose.[54] Dante's *Commedia* elaborates on the sixth book of the *Aeneid* in the *Inferno* in making Dante a type of Aeneas and Virgil a type of the Sybil.

Further, the "heroes" of these medieval "epics" can be recognized as fitting the definition of the epic's protagonist—as defined by the very medieval and allegorical Fulgentius, who wrote commentaries on the epics of Statius, Virgil, and Ovid, and whose setting in the *Mythologiae* is indebted to Martianus Capella.[55] In his commentary on the *Aeneid*, Fulgentius analyzed the epic hero Aeneas as representing the human ideal of "manliness of body" and "wisdom of mind" perfected in the *fortitudo* and *sapientia* of Christ;[56] he accordingly understands the *Aeneid* as an allegory of the ideal man's development from birth to maturity. By extension, the ideal epic hero of Martianus might be a combination of the two chief figures, the god Mercury, who represents eloquence, and the mortal Philology, who represents knowledge; their wedding symbolizes the human ideal as a combination of corporeal skill and spiritual or rational excellence. By extension, embodying the same Fulgentian definition, the ideal epic hero of Boethius' *Consolatio* can also be seen as fragmented into separate figures, represented by Philosophy (the rational faculty) and the narrator Boethius (the irrational earthbound "body"). Finally, the epic hero of the *Commedia* is represented by the character Dante, in combination with the guide Virgil (in the *Inferno* and *Purgatorio*) and Beatrice (in the *Paradiso*). Dante and Virgil signify, according to Dante's son Pietro Alighieri, "ratio," or "philosophia rationalis," and Dante and Beatrice, "studia theologico," or "theologia."[57] Behind these allegorical projections of the human ideal looms Christ, whose superhuman love for mankind is suggested by the Good Friday setting and by the thirty-three cantos of each cantica (thirty-four in the *Inferno*, although the first serves as an introduction to the whole of the *Commedia*). The numbers are significant because thirty-three was Christ's age at his death, and one hundred—for the total number of cantos—is a perfect number.

Why a medieval Christian studied Virgil and Martianus Capella is clear only after looking at the history of education from antiquity through the Middle Ages. The history of education, which relied on the classics, provides an important context for the history of mythography, or the allegorization of classical mythology; the high points of each are the same. Thus, in the Middle Ages the strongest revivals, or renaissances, of classical learning occurred during the eighth to tenth centuries, that is, the periods of Benedictine and Carolingian reform in England and on the continent; during the twelfth century, that is, the time of the triumph of the schools of Paris and Chartres; and during the fourteenth and fifteenth centuries, that is, the advent of the Italian and early English renaissances. And within these renaissances of classical learning so important for medieval education and mythography, different classical authors were taught and glossed in the schools—Virgil, Lucan, and Statius initially in the late antique period, supplemented by Boethius, Martianus Capella, and Fulgentius, in particular, in the Carolingian period, and then all of these plus Ovid and mythographic handbooks in the twelfth and fourteenth centuries, followed by the addition of Seneca and Dante in Italy. Such revivals of learning traced their inspiration back to the classical schools, which regarded the Greek Homer and the Roman Virgil as having almost divine authority, especially as masters of grammar and rhetoric, and therefore necessary for the study of the *trivium*.[58]

In the Roman schools, of course, young men destined for public life received an education bound up in part with "paganism," both because the rhetoricians and philosophers in these schools were pagan[59] and because allegorical interpretation helped defend the established polytheistic religions. Indeed, in the year 362, with Julian the Apostate's creation of the Scholastic Law, Christians who taught Homer and Hesiod but who did not believe in their gods were compelled to give up teaching.[60] Thus, both pagan and Christian children read Homer and the poets in the fourth-century schools, such study having been granted an immunity by Julian from the threats of Church and parent.

It should come as no surprise that at this time Virgil was used widely in the schools and received much attention from commentators and lecturers, particularly from North Africa. One of the chief proponents of Virgil was the North African scholar Macrobius. Like Donatus (whose commentaries are lost) and Servius in their commentaries on all of Virgil's works at the end of the fourth century, Macrobius "commented" on all aspects of Virgil's knowledge of the arts and related subjects in his *Saturnalia* in the early fifth century, and he glossed significant quotations in his commentary on Cicero's *Somnium Scipionis*. That Macrobius was African was no accident, nor was the fact of his commentaries on ancient classical works and culture. In the late Roman empire there grew up a close commercial and intellectual connection between Rome and its provinces in Africa, Spain, and Gaul, with the provinces eventually becoming more important than Italy (the Emperor Theodosius himself was from Spain).[61] In fact, African rhetoricians commenting on Virgil included Macrobius, Fulgentius, Martianus Capella, and Isidore of Seville; what they shared was an interest in etymological allegorism, the handbook and encyclopedia, and the seven liberal arts (Martianus' encyclopedic *De nuptiis* details the seven liberal arts in seven of its nine books—"perhaps the most widely used school book of the Middle Ages"[62]—and Isidore's friend posthumously organized the divisions of his encyclopedia *Etymologiae* roughly into divisions approximating the *artes*).

During the decline of Roman civilization Boethius' *Consolatio Philosophiae* became very popular as a schoolbook; the work's mythological allusions were explained by generations of commentators. The reason for Boethius' importance stemmed from this age's great interest, from this point forward until the twelfth century, in education and the *artes*, which mirrored Lady Philosophy's advocation of the pursuit of wisdom to the desolate Boethius.

Roman civilization disappeared from Africa first, but lingered on in Spain, where the classical writers continued to be read;[63] with the Germanic invasions of the fifth century, Roman civilization was destroyed, except perhaps in Vandal Africa and Italy. Civilization was not destroyed in the latter perhaps because of the humanistic interests of the uneducated Ostrogoth king, Theodoric (493–526), under whom Boethius served.[64] Theodoric's connections with both Boethius and Cassiodorus deeply influenced the history of mythography at this time. The Ostrogoth attracted to him not only Boethius but also Cassiodorus, who was appointed *magister officiorum*, chief of the chancellery, in 523 and who instituted various educational reforms—he brought students to

Rome from the provinces; built Vivarium, a double monastery in Calabria with a vast library that attempted to synthesize traditional humanism with Christianity; and wrote the two-volumed *Institutiones* outlining his system of education.

Although education had remained dual, with the secular adhering faithfully to pagan humanism and the clerical to religious asceticism,[65] the arrival of the Germanic Lombards in 568 and their conquest of Italy in 572 brought to most of the country the barbarism that had devastated the rest of Europe since the early fifth century and expunged secular education. Only religious education remained available, primarily to ensure an adequate supply of educated priests, either in the monasteries and their schools (with the influence of Celtic monasticism now being felt on the continent) or in episcopal schools (as occurred very early in Gaul).[66] Secular education, when it occurred at all, took place in the home in the Merovingian age of the seventh century, as is evidenced in the example of Felix of Pavia educating his own nephew Flavian, who later became tutor to Paul the Deacon.[67]

A more traditionally classical form of education did continue in England and Ireland in the late sixth and seventh centuries because of the missionaries sent from Italy—the first, St. Augustine, sent by St. Gregory the Great in 597 and later named first Archbishop of Canterbury, and the second, his Greek successor, Theodore of Tarsus. Benedict Biscop stocked the famous monasteries Wearmouth and Jarrow in the north of England with books collected from six different trips to Rome.[68] In addition, in the early Irish schools study of the classics, especially Greek, continued,[69] perhaps because their priests came from Britain, Gaul, and later, Spain. Only in Ireland, where Christian culture was not so burdened with classical associations and where the pre-Christian native culture had always been tolerated by the Church, could the schools synthesize Christianity and pagan antiquity and send it to Northern England and Scotland via Irish monasticism and art.[70]

Among these teachers and missionaries might have been the First Vatican Mythographer,[71] possibly the same Adanan the Scot of the eighth century who glossed the *Eclogues* and *Georgics* of Virgil in an eccentric allegorical way through elaborate political parallels and who brought to Hibernian studies an interest in Greek and Roman myth, gathering loosely into one handbook all the myths he could obtain from Hyginus and from Lactantius Placidus' commentary on Statius. He is important as perhaps the first medieval scholar who, in his desire to collect and assimilate myths, can be termed "mythographer" rather than merely "commentator."

The Irish schools of the Merovingian period inspired renaissances both in the eighth-century Benedictine age of Northumbria and in the ninth- to tenth-century Carolingian age of France; these renaissances would lead to gloss and commentary on Martianus Capella and Boethius and to extensive use of Fulgentius, in particular. The classical tradition reemerged in a new medieval cultural synthesis of Lombard and British cultural elements exemplified by the meeting of the English Alcuin and the Lombard Paulus Diaconus (Paul the Deacon) at the court of Charlemagne.[72] Indeed, the spur of the Carolingian revival of classical humanism was the Irish and Northumbrian cleric: this scholar was involved in the creation by Charlemagne of the great court schools and in the creation by the Scoti (or the Irish, called so until the eleventh century)[73] of monastic schools and missions on the continent as well as in copying of classical texts, writing of commentaries, and book-making. It was the Irish scholar or cleric and not the monk or priest who departed for the continent primarily to revive the learning that had been diminished by the Norse.[74] This Irish cleric's traces have been found in northern France, Burgundy, the territories of modern Switzerland and northern Italy, the Rhine Valley, Franconia, Bavaria, and the Salzburg area.[75] In addition to preserving the early Irish language[76] the cleric also preserved Latin and education in general:

> There is a line from Ireland and Iona to Jarrow and York, and from there to the Court of Charles. Alcuin's school at Tours is the parent of the school at Fulda where Hraban carried on the same work. Different lines of descent are united at Reichenau and St. Gall, which are in relation with the newer school at Fulda on the one hand, and with the Irish on the other. Bede (Jarrow) taught Egbert (York), who taught Alcuin (Tours), who taught Hraban (Fulda), who taught Walafrid Strabo (Reichenau). . . .[77]

Of the activities of the Irish cleric mentioned above, the most significant for the history of mythography was the writing of commentaries. The Scoti commented on both Martianus Capella and Boethius, probably because both were used as school authors. The earliest extant (and not very important) glosses derive from the ninth century and may have been lecture notes of teachers or their transcription by students.[78] These early glosses on Martianus were expanded by Dunchad (or possibly Martin of Laon) and by John Scot, who then catalyzed the writing of the more famous and influential late ninth-century commentary by Remigius of Auxerre on the continent.

In addition to these ninth-century commentators there also existed influential mythographers who lent credence to that title by themselves collecting myths and moralizing them. Of these, Remigius was the single most important figure: he can probably be identified as the Second Vatican Mythographer, or more precisely, the figure responsible for the reorganization of the myths initially compiled by the First Vatican Mythographer. This identification makes sense, given the echoes of mythographic (chiefly Fulgentian) material from Remigius' Boethius and Martianus commentaries also found in the Second Vatican Mythographer; Remigius also wrote commentaries on Terence, Priscian, and the *Aeneid*. Other mythographers of this period included the more minor Spanish bishop Theodulf, who taught at Orleans and wrote a poem with mythological elements, "De libris quos legere," in imitation of Ovid's *Metamorphoses* and indebted to Isidore's *Etymologiae*; and the author of the very important *Ecloga Theoduli* (possibly Gottschalk), who debated in this heavily mythographic work the superiority of biblical verses to classical stories.

In the tenth and eleventh centuries a lull occurred in the development of humanistic studies, although the Carolingian interest in certain of the classics, especially Martianus Capella and Boethius, continued, if in a literalistic and diminished way. Much of the mythographic work became derivative, minor, and performed by scholars in geographical regions outside the major schools of the Carolingian and the twelfth-century renaissances. In addition, translations, grammatical glosses, and dictionaries predominated among the types of commentaries extant. In Italy in the tenth century Rather of Verona provided minimal glosses on Martianus Capella, and in the eleventh century Papias the Lombard completed his dictionary indebted, in part, to the Carolingian commentaries on Martianus Capella with their plethora of references to obscure Roman gods. In Germany a similar interest was furthered by Notker Labeo, who translated both Martianus' *De nuptiis* and Boethius' *Consolatio* into Old High German, in the process transcribing many of Remigius' glosses.

The only new development in this rather sterile period of the tenth and eleventh centuries stemmed from the grammatical interest in Ovid, Lucan, and Statius: literalistic glosses on Ovid, supposedly notes from the eleventh-century lectures of the German Manegold of Lautenbach on the *Metamorphoses*, appear in Benediktbeuern manuscript 4610 and also in minor anonymous commentaries on Ovid's *Fasti* from England and Brussels; there are various tenth- and eleventh-century scholia on Lucan, mostly from Switzerland and France; scholia on Statius' *Achilleid* were written.

Two relatively rich sources of mythographic material—and certainly reflecting the earliest explicit interest in either Fulgentius or the *Ecloga Theoduli*, which at this time came to be used as a schoolbook—were the eleventh-century fragment by the French Baudri de Bourgueil of a moralized mythology taken verbatim from the *Mythologiae* of Fulgentius and the late eleventh-century, very full commentary by Bernard of

Utrecht on the *Ecloga*. This interest in Lucan, Ovid, Fulgentius, and the
Ecloga—in addition to the elaboration of the earlier Carolingian interest in Boethius and Martianus and the interest in Virgil last expressed by
the late antique commentators—began to grow in the twelfth century.

It was also in the twelfth century that the high-water mark in medieval commentary on Boethius, Martianus, Virgil, and Ovid was reached
via the intensified and intense interest in the classics found especially
in France and England, where the focus of study shifted from the rhetoric and grammar of times past to dialectic and philosophy.[79] Boethius
commentators looked for Neoplatonic parallels in the *Consolatio*, especially in the ninth meter of the third book, beginning "O qui perpetua";
the various mythic portions were treated similarly by the Erfurt Commentator and the very important Chartrian philosopher William of
Conches, the latter of whom also wrote explicitly Neoplatonic commentaries on the *Timaeus* and on Macrobius' commentary on the *Somnium Scipionis*. Martianus commentaries continued to be written,
among them one by the Anonymous Barberinus (whose commentary is
an expansion of the earlier Remigius), an anonymous Florentine commentary, a fragmentary commentary on the first book by Bernard Silvester, and a long, fairly important one by the English Alexander Neckam.
These commentaries focused only on the first two (allegorical) books of
De nuptiis, which introduce the remaining seven on the liberal arts.
The Fulgentian view of the *Aeneid* as an allegory of human life was
elaborated by Bernard Silvester in his extensive commentary on the first
six books, especially the sixth, and adopted by the English John Salisbury in his various comments in the *Policraticus*.

Other interests from the tenth and eleventh centuries burgeoned with
the attention to the *integumentum* of classical myth as it "cloaks"
hidden truth, mostly moral or cosmological (Neoplatonic).[80] There are
moralized commentaries on Lucan and Ovid (on both *Fasti* and the *Metamorphoses*) by the influential Arnulf of Orleans and Fulgentian handbooks by the English Digby Mythographer and the extremely important
Third Vatican Mythographer (who can be identified either as Alexander
Neckam or Albericus of London and who wrote a well-organized allegorical version of the myths contained in the handbooks of the first two
Vatican Mythographers). The dictionary tradition begun by Papias continued, with many mythological references, in etymological works by
Osbern of Gloucester and Hugutio of Pisa. There is also a commentary
on the *Ecloga Theoduli* by Alexander Neckam, which became so popular that it drove out that of Bernard of Utrecht.

The reasons for this particular twelfth-century renaissance have been
well documented. In the twelfth century the French cathedral schools,

especially in Chartres, Orleans, Rheims, and Laon, had continued the work of the Carolingian humanists, combining study of the best of the ancients and the Church Fathers; from the ninth to the eleventh century they had trained men for the Church by giving them a very rigorous education, unlike the monastery schools where standards had fallen and secular studies had been discouraged.[81] In the early twelfth century a change occurred: at the episcopal, or chapter, schools associated with cathedrals, like Chartres, where the bishop had always been the chief teacher, the bishop became too busy (or powerful) and thus delegated his functions to an assistant, who came to be known as the *Magister scholae*. [82] Thus, in France and Germany more than in Italy the Latin classics were revived and the libraries expanded. The flurry of academic activity resulted in one quarter of the *Patrologia latina* being written during this period.[83] Most twelfth-century libraries (as we can tell from the catalogs for particular libraries then and in later centuries)[84] included some classics, usually at least the logical works of Aristotle and those of Virgil as a school author, although Greek books generally were rare.[85] Poets were read more than prose-writers, except for textbook authors like Martianus and Boethius, although the range of authors read had shrunk from Carolingian times, partly because of the development of *florilegia*. These anthologies of citations from individual authors came to be used in place of the entire texts, a phenomenon that probably originated in northern France.[86] The resurgence of interest in the classics and the growth of the cathedral schools at the beginning of the twelfth century led to the development by the end of the century of the first universities at Salerno, Bologna, Paris, Montpellier, and Oxford.[87]

In the thirteenth century the interest in the classical epic continued but, as in the watershed tenth and eleventh centuries, in a subdued fashion. Elementary education as always focused on the classics, although Ovid and Statius were not included among authors read,[88] but very few commentaries with mythographic significance were written at this time. There were several reasons for this: interest had shifted from literature to philosophy and science, an interest stimulated by the translations from Greek and Arabic and the rediscovery of Aristotle through Averroes;[89] further, the writings of the twelfth-century masters drove out the classics in the thirteenth, largely because these modern writers were themselves strongly indebted to the classics and the humanistic tradition.[90] Thus, only two relatively minor texts written in the thirteenth century contributed to the mythographic tradition—John of Garland's *Integumentii Ovidii* and Giovanni Balbi's dictionary, heavily indebted to his predecessors Papias, Osbern, and especially Hugutio. The one development of the period strongly influencing subsequent centuries was the spread of classical *florilegia*, or excerpts, supposedly responsible for encouraging the humanism that flowered in the sixteenth-century English Renaissance.[91]

The major interests of French and English mythographers of the fourteenth and fifteenth centuries centered on Ovid and Boethius, with the earlier interest in Virgil, Lucan, and Martianus almost dying out and being replaced by a new interest in Seneca, Augustine, and the *Ecloga Theoduli*, and by new mythographic techniques. Of these phenomena, the most important involved Ovid. Ovidian commentaries and handbooks became longer, more elaborate, and even more heavily allegorical, especially the very important *Ovidius Moralizatus* of the French Pierre Bersuire. This "moralized Ovid" derived from the epitome of the *Metamorphoses* in the *Narrationes* of Lactantius Placidus, which was later moralized and elaborated by Arnulf of Orleans, and from the immense compendium of the *Ovide Moralisé*, in French verse, transcribed into prose and published by Colard Mansion in the fifteenth century. In England the major Ovidian and Fulgentian interests of mythographers were reflected in the important *Fulgentius metaforalis* of John Ridewall; the *De deorum imaginibus libellus*, an anonymous Latin handbook using Bersuire's early draft of the *Ovidius Moralizatus* as a base; and the fifteenth-century *De archana deorum* of Thomas Walsingham, among others.

In addition, a plethora of Boethius commentaries introduced Aristotelianism into their allegorical interpretations. The Boethius commentators ranged from William of Aragon, Tholomaeus de Asinariis, the False Thomas Aquinas, Pierre D'Ailly, Regnier of St. Tron, and Dionysius Cartusianus to the fifteenth-century Arnoul Greban and Josse Bade d'Assche in France, with Nicholas Trivet in the early fourteenth century in England and Guillermus de Cortumelia in Spain.

The most important of these commentators, Nicholas Trivet, anticipated the Italian and French renaissances with his commentary on Seneca's tragedies and, apparently in tandem with Thomas Waleys, with a commentary on Augustine's *De civitate Dei*. There also appeared commentaries on Augustine by John Ridewall and John Baconthorpe. In the same period the *Ecloga Theoduli* received new attention from a commentator in England known as Steve Patrington, who depended heavily on Alexander Neckam's earlier commentary, from Odo Picardus in France, and in the fifteenth century from the Anonymous Teutonicus in Germany. New mythographic techniques developed too. Special *exempla* called "pictures," drawn from classical mythology and moralized, adorned the homiletic and patristic writings of fourteenth-century friars like Nicholas Trivet, Robert Holkot, and Thomas Hopeman.[92]

Also at this time in Italy Ovidian handbooks and commentaries flourished along with Dante commentaries written to explain unusual terms and mythological figures in the *Commedia*. Specific commentaries on the *Metamorphoses* included the allegorical commentary by Giovanni del Virgilio and the moralized translation by Giovanni Bonsignore. Mythographic encyclopedias designed to present complete and exhaustive material about the gods provided, in addition, defenses of poetry and theories of various types of poetry. These manuals, whose authors returned to classical sources for information rather than relying on medieval *florilegia* culled from those sources, included two very important works, the *Genealogie deorum gentilium libri* of Boccaccio and the fifteenth-century *De laboribus Herculis* of Coluccio Salutati. It is possible Salutati was influenced by the fourteenth-century Trivet's commentaries on Seneca's Hercules tragedies, or those of his contemporary Luca Mannelli, or the fifteenth-century Bernardinus Marmita and Danielus Galentanus; probably he himself influenced the fifteenth-century Spanish work on Hercules by Henry of Aragon.

In addition to the Ovidian commentaries and Fulgentian handbooks, there also appeared primarily grammatical commentaries on Virgil and Lucan and a variety of commentaries on Dante. Among those glossing the *Aeneid* we find Petrarch, Benvenuto Rambaldi da Imola, Folchino Borfoni, Pomponio Leto, Zono de Magnalis, and the important Cristoforo Landino, all of whom except Landino also wrote commentaries on Lucan. Finally, various fourteenth- and fifteenth-century writers, many of whom borrowed material from one another, commented on Dante. The most allegorical commentaries belong to Jacopo della Lana, Pietro Alighieri, Giovanni Boccaccio, and to an extent Benvenuto Rambaldi da Imola. The others, from the earliest to the latest in existence, inform, describe, and explain rather than allegorize: the Ottimo commentary, the Anonymous Selmi, those of Graziolo de Bambaglioli, Jacopo Alighieri, Fra'Guido da Pisa, the so-called False Boccaccio and Francesco da Buti, the Anonymous Florentine, and in the fifteenth century, those of Giovanni da Serravalle, Guiniforto delli Bargigi, Stefano Talice da Ricaldone, and Cristoforo Landino. It is perhaps appropriate, then, for us to look at Dante ourselves to determine how this late medieval poet who was himself so commented on by succeeding scholars used the preceding mythographic commentaries to reinterpret Virgil and his epic hero Aeneas. In the process we shall analyze with greater particularity the way mythography actually works.

III

To explain how mythography functioned in the Middle Ages we shall look at Dante as a mythographer who drew upon earlier mythographers. The example we shall use involves his attempt to Christianize one of the greatest classical poems—Virgil's epic, the *Aeneid*. While Dante deliberately modeled his own "epic" on Virgil's, in that he expanded the sixth book of the *Aeneid*, on Aeneas' descent into the Underworld, into a frame for the *Inferno*, he also modeled his Christian character Dante on the epic hero Aeneas, who made a similar extended journey throughout the world, and he selected Virgil as his protagonist's guide and teacher in the first two *cantiche* of the *Commedia*. What we shall probe in this investigation is the mythographic reasons for some of these choices.

Dante viewed Aeneas as a human ideal. In his *Il Convivio*, or *The Banquet*, Aeneas epitomizes spiritual and national piety. That is, his piety is said to be his greatest praise; he is the "light" and "hope" of his country; he represents perfect Youth.[93] As perfect Youth, Aeneas epitomizes the ideal second stage, out of four, in man's life: Dante presents an allegorical scheme of human life (taken from Albertus Magnus) as having four stages, Adolescence, Youth, Old Age, and Decline.[94] These four stages are related to four combinations of the humors, probably based on a theory of astrological types (Adolescence, warm and moist; Youth, warm and dry; Old Age, cold and dry; Decline, cold and moist), the four seasons, the four parts of the day, as it was divided by the Church, and the four horses of the sun's chariot.

The reason that Aeneas represents ideal Youth is demonstrated by his change from passionate to brave in books Four to Six of the *Aeneid*. Aeneas, declares Dante, was unbridled in youth because of the pleasure he derived from his liaison with Dido, in Book Four, but he acquired the virtue of fortitude in order to enter Hades with the Sibyl and together seek the soul of his father in Book Six. Dante's example concludes with a moral admonition: be brave and temperate in youth, like Aeneas, so that Reason controls Desire just as a rider controls his horse (*Convivio*, 4.26, pp. 284–85). But Aeneas manifests other virtues as well. He is praised for his love of his elders and juniors, for example, when he left behind the aged Trojans in Italy and when he instructed Ascanius and his son and the other youths in the tournament later (Book Six, *Convivio* 4.26, p. 285), for his courtesy, another necessary virtue of Youth, when he honored the corpse of Misenus by chopping logs for the funeral pyre (Book Six, *Convivio* 4.26, p. 285), and for his loyalty, when he scheduled the Sicilian games on the anniversary of Anchises' death (Book Five, *Convivio* 4.26, p. 287).

For Petrarch, Dante's contemporary, Aeneas is also the Perfect Man whose heroic exploits illustrate the rational man's triumph of Virtue over Vice (specifically, passion). In a Latin letter about the fictions of Virgil, "De quibusdam fictionibus Virgilii,"[95] Petrarch declares that "Virgil's subject, as I understand the matter, is the Perfect Man." But most of his comments in this letter and elsewhere on the *Aeneid* center on Aeneas' problems with Vice rather than the triumph of his Virtue. Petrarch interprets the voyages of Aeneas as a moralized figure for man's spiritual dilemmas. The Fall of Troy in particular epitomizes a descent into vice—passion, as represented by Aeneas' desire for battle, the influence on him of his mother Venus, and his association with the drunken old Trojans who do not anticipate the final attack of the Greeks. Venus is described by Petrarch as a goddess

> . . . whose pursuit by us becomes hotter and keener *toward the middle of our life*. Her assumption of a maidenly look and air is for the purpose of deceiving the unwary. If we saw her as she is we should flee from her in fear and trembling for, as there is nothing more tempting than pleasure, so there is nothing more foul. Her garments are girded up because her flight is swift. For this reason she is compared to the swiftest of creatures and things. It cannot be denied that nothing swifter exists, whether you consider her comprehensively or part by part; for pleasure as a whole passes from us very soon, and even while it still abides with us each taste of it lasts but a moment. And then, finally, she appears in the garb of a huntress, because she hunts for the souls of miserable mortals, and she has a bow, and has flowing hair, in order that she may smite us and charm us.

A similar view of Venus' vicious influence is provided by Petrarch's *Secretum meum*, a three-part dialogue on the theme of *contemptus mundi*, "contempt for the world." Petrarch explains how Venus as passion clouded Aeneas' sight with love of the things of this earth, which he could see truly and more clearly only when her influence ended—a moralization that refers to Aeneas' passionate desire for battle in the Fall of Troy, at a time when he is still guided by his mother Venus (in *Aeneid* 2.361–9 and 2.262).[96] Further, in the Third Dialogue Petrarch interprets figuratively the imminent fall of Troy—the city with which she was associated in the *Aeneid*—as vicious. The Trojans do not worry about death because they are drunk with temporal pleasure; Petrarch's character, Augustine, relates this to man's viciousness in old age (again, note the relationship with a stage in life reminiscent of Dante's schema). When Petrarch the character asks Augustine why man wants to stretch out his short life, Augustine responds that one cause is man's love of pleasure. He then provides an elaborate analogy between the old Trojans (representative of human nature) and the Fall of Troy (fall into viciousness through desire for pleasure):

> . . . you will persist in letting old age find you still in the midst of games and empty pleasures; like the old Trojans who in their customary ways passed the last night without perceiving
>
> > *The cunning, fatal horse, who bore within*
> > *Those armed bands, had overleapt the wall*
> > *Of Pergamos.* [*Aeneid* 6.515–6]
>
> Yes, even so you perceive not that old age, bringing in his train the armed warrior Death, unpitying and stern, has over-leapt the weakly guarded rampart of your body; and then you find your foe has already glided by stealth along his rope—
>
> > *And now the invader climbs within the gate*
> > *And takes the city in its drunken sleep.* [*Aeneid* 2.265]
>
> For in the gross body and the pleasure of things temporal, not less drunk are you than those old Trojans were, as Virgil saw them, in their slumber and their wine. (*Secretum meum*, pp. 158–59.)

It is interesting to note that even though Aeneas is not explicitly mentioned in this passage, the *Aeneid* (Books Two and Six) itself is glossed in a discussion of old age, one of Dante's four divisions of human life.

According to Petrarch, Aeneas' evolution into the Perfect Man is aided by the attacks of Aeolus, the god of winds who in the poet's "Letter on Virgil's Fictions" represents "our reason, which curbs and controls these headstrong passions." In the *Secretum meum*, as well, Petrarch explains how Aeolus represents the rational soul lodged in man's head, which controls anger in the heart and desire in the loins. Aeolus, as god of winds, attempted to scatter the fugitive Aeneas and his fleeing Trojans at the request of Juno, whose pride still smarted over the Judgment of Paris, when Venus was selected as most beautiful instead of Juno or Minerva, and who was fearful over the prophecy that men of Trojan blood would someday ravage her city, Carthage. For Petrarch in the Letter and in the *Secretum meum*, this god of winds sits above mountains piled on top of deep caves to show that the home of reason, the head, is literally housed above angry passions.

Throughout Dante's and Petrarch's moralization of Aeneas as the Perfect Man or Youth, the poets have understood the Mediterranean landscape as a metaphor for man's psyche and the hero himself as a type of Everyman torn between passion and reason. Aeolus' home is the head; Venus', presumably, is the loins. Indeed, Petrarch interprets the cosmos as the microcosm man in commenting on the Stoic concept of the World Soul as fiery seed animating and generating all the universe (in *Aeneid* 6.730): "In effect, he [Virgil] has given us to understand he means by the earth our bodily frame; by the sea, the water through which it lives; and by the depths of the sky, the soul that has its dwelling in a place remote, and of which elsewhere he says that its essence is formed out of a divine fire" (*Secretum meum*, p. 101). The questions we now want to ask are, how do the hero, the Trojans, and these gods, Venus and Aeolus, come to be moralized in this curious way, and, second, why are the *Aeneid* and its characters linked with the stages of human life?

These views of the gods and heroes and of the *Aeneid* derive from earlier medieval glosses and commentaries on the *Aeneid*. Such commentaries included those of the fourth-century Servius, the sixth-century Fulgentius, the anonymous late tenth-century glossator on the sixth book of the *Aeneid*, possibly Remigius of Auxerre, and the twelfth-century commentary on the first six books of the *Aeneid* by Bernard Silvester. A brief look at the most influential commentaries for Dante and Petrarch—particularly those of Fulgentius, Bernard and his contemporary William of Conches writing about Boethius, and Remigius of Auxerre—will illuminate these Italian poets' moralized understanding of Virgil's *Aeneid*.

The view of Aeneas as a Virtuous or Rational Man depends on an interpretation of the *Aeneid* as an allegory of human life. The first mythographer to interpret the epic in this way was Fulgentius in his sixth-century *Expositio continentiae Virgilianae secundum philosophos moralis*, like Petrarch's *Secretum meum* a dialogue between its author as character and a *magister* or teacher, in this case Virgil himself rather than St. Augustine. Here Aeneas perfectly combines *virtus* as manliness and wisdom as he progresses through the twelve books of the epic, understood by Fulgentius as representative of the stages of human life—birth, infancy, adolescence, passion and youth in the first five books, learning and education in the sixth book, and experience (or the increase of good through virtue) in the seventh to twelfth books.

In the twelfth century Bernard Silvester next adapted Fulgentius' allegorical frame to unify his own commentary on the *Aeneid*, itself heavily indebted to specific Servian glosses on that epic, but Bernard's commentary stopped after six books, and he made each of these six books representative of one stage of life—infancy, boyhood, adolescence, youth, manhood, and the afterlife. The most attention is devoted to the sixth book, centering on Aeneas' descent into the Underworld, or Hades, the latter understood as a metaphor itself for man's life on earth. Thus, Petrarch's later understanding of the cosmos as an allegorical symbol for the microcosm has an antecedent in Bernard's own comments on that same sixth book on which Petrarch's gloss is based. For Bernard, Aeneas represents the human spirit occupying the body in that the name Aeneas comes from *ennos demas,* "occupant of the body."[97] Because the human body is a "lower thing" and nothing is lower than such *inferos,* the human body becomes an *infernum,* or Underworld.[98] Hence, man's life on earth can be seen as a "descent" into an Underworld. Man descends into this Underworld in four ways, Bernard explains in the preface to the sixth book: naturally, at birth; viciously, when he succumbs to temporal pleasure; virtuously, when he arms himself with knowledge of the Underworld in order to protect himself from its temptations; and finally, artificially, when he consults with demons about the afterlife (6.1, Meerson, pp. 128–29; Jones, p. 30).

Aeneas' descent to Hades to find the spirit of his father is, according to Bernard's definition, clearly an artificial descent. But it can also be interpreted as a virtuous descent, as it was by Fulgentius, when it is viewed as an allegory of Aeneas' education in the seven liberal arts—the *trivium* and *quadrivium*—to protect him from the temptations of this "Underworld." Further, according to Bernard, other heroes like Aeneas have made similar virtuous descents—Orpheus, Theseus, and Hercules (6.56 and 6.87–9). In each case the descent involves some contact with vice or concupiscence, just as Petrarch's Aeneas was involved with figures like Venus and the Trojans who represented vice. Orpheus is a type of the wise and eloquent man who descends into the Underworld to retrieve his wife, Euridice, or concupiscence, who has been poisoned by the snakebite of pleasure in temporal good. Theseus is the "rational and virtuous man who descends into the Underworld in accordance with the virtuous descent" (Meerson, p. 157). Hercules also descends virtuously and is able to return because he is semi-divine and because he "dragged forth the defeated door-keeper Cerberus. Hercules symbolizes victory over vices. . . . He descends into the Underworld when he comes through contemplation to temporal things, but because he is semi-divine (spiritually rational and immortal; corporeally irrational and mortal), he turns away from them when he rises again to celestial things" (6.392, Meerson, p. 191). Hercules is also discussed as a wise man who battles the Hydra of ignorance (6.392, Meerson, p. 191). Finally, Theseus and Pirithous are dear friends who represent the wise man and the eloquent man respectively; they descend into the Underworld to rape Proserpina. "This means that when some men come to know the course of the moon and the other similar natures of the stars, they love to play the philosopher about mundane things; the wise man, however, is led forth by his own wisdom. The man who is solely eloquent is completely conquered by his own garrulity" (6.393, Meerson, p. 193).

Although Bernard derived the concept of the four descents from William of Conches' twelfth-century glosses on Orpheus in Boethius' *Consolatio Philosophiae,* he applies the system here to the *Aeneid* in a fairly original way, perhaps inspired by Fulgentius' own treatment of the sixth book as an allegory of education in the arts. In addition, his specific application of the vicious and virtuous descents to the *Aeneid* may represent an elaboration of Servius' comments on the golden bough in the sixth book (6.136, 6.295, and 6.477), which recur not only in Bernard but also in a late tenth-century gloss on this sixth book. In Servius' commentary he compares the golden bough with the letter Y, viewed by Pythagoras as a symbol of human life. Man, at the time of adolescence, must follow either the path on the left, that of vice, or the path on the right, that of virtue (just as, in Virgil's description of the orthography of the lower world, Aeneas' path through Hades branches off to the right to Tartarus, place of the wicked [6.295] and then to the left, to Elysium and the place of the virtuous [6.477]). In Bernard's commentary the use of the golden bough becomes the Wise Man's arming of himself in order to descend into the Underworld.

> Intelligence advises him to seek this branch so that he can open the passage to the Underworld; if someone lacks philosophy, he does not have access to real knowledge about things. The right branch is raised, signifying the entry to life, while death is signified by the depressed left branch, meaning the point of entrance into substance from non-being. The branches are conjoined in a crook, disjoined in discreteness. This bough is in a TREE. Pythagoras called humanity a tree, seeing that it divides itself into two branches, namely virtue and vice. For though it is whole at first, later it divides itself, part to the right and part to the left, that is, some into vice and some into virtue. Furthermore, this tree is THICKLY SHADED by the weightiness of the flesh. So here humanity is called a tree because it is divided like one; Pythagoras represents it by the symbol Y, the character which has the form of a forked tree (6.58, Meerson, p. 160).

Hence the Y as a symbol of human life with its two paths, virtuous and vicious, may have prompted Bernard to apply William's system of descents to the sixth book of the *Aeneid.* This Pythagorean Y was also used to classify and distinguish the nine circles of hell in a late tenth-century gloss on the sixth book of the *Aeneid,* which may also have strengthened Bernard's equation of the two paths in life with the two descents into the Underworld, vicious and virtuous. In Parisinus 7930, which also contains the *Eclogues, Georgics,* and *Aeneid,* with marginal and textual annotations, the anonymous author (possibly Remigius of Auxerre) cites Servius' notes on the circles of hell (6.426, 430, 434, 440, 477, 548, 638, 656, 679) and on the golden bough and the Pythagorean Y in structuring his diagram of the Y. There is one circle attached to the lowest part of the Y, two on each side of the lower limb, and similarly for the upper part there are two circles on the right and left of the upper two limbs.[99] The captions for the various circles divide the inhabitants into those who are saved and damned, in body and/or in soul. Given Servius' identification of the nine spheres or circles of the cosmos (the seven planetary spheres plus two other "great circles") with the nine circuits of the Styx in 6.127, to reveal that Virgil was cloaking truth with poetic fiction in describing Hades in the sixth book, the Parisinus glossator has firmly clinched the connection between this world and the Underworld, between moral choice in life as represented by the two paths of the Pythagorean Y and the two descents, virtuous and vicious, into the Underworld.

That Dante and Petrarch were familiar with the tradition of medieval descents can be seen in their treatment of Virgil, if not of Aeneas himself. Dante, of course, used Virgil as a guide for his character Dante in the *Commedia,* an apt instructor in the nature and inhabitants of the Inferno and Purgatorio. Petrarch also characterizes Virgil as a Wise Man similar in his wisdom to his epic hero Aeneas in his Latin verses on the *Aeneid,* "Ad publium Virgilium Maronem heroycum poetam et latinorum principem poetarum." He asks Virgil, confident of the latter's familiarity with the cosmos and comparable, therefore, to Aeneas in his descent into the Underworld, where he abides now: "What region of earth or what circle of Avernus arrests thee now?"[100] Petrarch goes so far as to compare Virgil, as the poet who plumbed the depths of the Underworld in his epic, with Christ, as the harrower of Hell.

> Wert thou received thither after the conquest of the Stygian abodes and the plundering of the Tartarean regions, on the arrival of the Highest King who, victorious in the great struggle, crossed the unholy threshold with pierced feet, and, irresistable, beat down the unyielding bars of hell with His pierced hands, and hurled its gates from their horrid-sounding hinges? ("Ad publium Virgilium Maronem," p.137)

In this letter to Virgil, Petrarch refers to other poet-heroes who also descended into the Underworld (Orpheus, line 14) and poet-pagan gods and places linked by Christians with the Underworld (swarthy Apollo with his "harsh and grating lyre," line 7, the "sable sisters," the Elysian groves and the Tartarian Helicon, lines 9–10). Perhaps he was familiar with Bernard's discussion of the heroes Orpheus, Theseus, and Hercules, all of whom had descended into the Underworld.

Thus Virgil, for Dante and Petrarch, represents a pagan poet whose own virtue and wisdom allowed him to explore the Underworld in the *Aeneid* in the manner of his own character Aeneas. And we also see Dante and Petrarch's Aeneas as a hero like Orpheus, Theseus, and Hercules. Like them, he was confronted by two choices, the path of Virtue or of Vice; like them, he initially succumbed to passion (Orpheus for Euridice and Theseus for Proserpina, but in Aeneas' case, substitute for Euridice or Proserpina either Venus' influence, the debilitating effect of the old Trojans, or the desire for battle). Like them, he becomes a wise man after being buffeted by reason as represented by Aeolus as god of winds, or the Perfect Man who can arm himself with virtue in order to descend into the Underworld in the sixth book. The emphasis of both Dante and Petrarch on the stages of life in their glosses on the *Aeneid* thus also derives from earlier moralizations and mythographized treatments of the epic, particularly the view of Aeneas as the perfect epitome of virtue and wisdom placed within the context of the "journey" of human life. Finally, both Dante and Petrarch emulated Virgil as an epic poet in creating their own epics, Dante in the *Commedia* and Petrarch in his unfinished Latin epic *Africa*, which "glosses" Cicero's *Somnium Scipionis* just as the *Commedia* "glosses" the *Aeneid*.

We have seen in this study that medieval mythography developed at least initially in a fashion parallel to that of classical mythography, which commenced with the defense of Homer's epic treatments of the gods and ultimately developed into more philosophical and allegorical treatments of the gods. That is, the mythographic tradition in the Middle Ages began with the classical or late classical schoolbooks, some of which happened to be epics retaining at least vestiges of the early Theophrastan definition that centered on gods, heroes, and men. In the latest medieval "epics," like the *Commedia*, "gods" ultimately became personifications and "heroes" became weak individuals who required divine assistance in their epic quests. In addition, the works glossed changed so that a relatively late schoolbook like the *Ecloga Theoduli* was not epic at all but a collection of eclogues, that is, short pastoral poems supposedly sung by shepherds in competition with one another and modeled on those eclogues of the epic-writer and great "magister" Virgil. Interestingly enough, in the fourteenth century, especially in Italy, the close relationship between epic and mythographic commen-

tary was revealed in the mixing of the two forms. Dante's epic can be
seen as a "commentary" on Virgil's *Aeneid,* especially the sixth book
with its descent into the Underworld. Similarly, Petrarch's Latin epic *Africa* (which includes a long catalog of gods in its third book) can be
viewed as a "gloss" on Macrobius' commentary on the *Somnium Scipionis* in that it relates the story of Scipio's conquest of Carthage at the
end of the second Punic War. Further, the mythographic manuals of Boccaccio and Salutati borrow epic themes or structures to unify their material. Boccaccio's *Genealogia* begins with an "epic hero" similar to Augustus, descendant of the gods in the *Metamorphoses*—only this hero is
the progenitor of the gods, Demogorgon. The name of this primeval creator, derived from a lost eleventh-century work by Theodontius, combines *daemon,* suggesting the infernal, with Gorgon, the terrible and
fierce quality associated with the gorgons. Because the pagan gods were
viewed as demons in the Middle Ages, the name of Demogorgon suggests that the reader of Boccaccio's *Genealogia,* like Dante in the *Commedia,* will explore Hell. Finally, Salutati's encyclopedia also uses the
sixth book of the *Aeneid* as a unifying device, for its fourth and last
book discusses various kinds of descents into Hell. Even the first two
books defend poetry, itself a descent into the *infernum* of artifice or fable, and the third book details the labors of Salutati's "epic hero,"
Hercules.

While the classical epic might relate the history of a city (Thebes in
the *Thebaid*), a nation (Italy in the *Aeneid*), or the world (in the *Metamorphoses*), or the genealogy of the hero (Augustus in the *Metamorphoses*), the medieval epic instead relates the spiritual "history" or story of
man as well as of the otherworlds of the gods and demons inherited from
both pagan and Christian literature. It is similarly clear that, if classical
mythography of the Greek and Roman poets explains literature to the
philosophers, then medieval mythography explains the gods and heroes
to the theologians—and to medieval poets like Dante.[101]

THREE

The Dimensions of the Discarded Image
Cosmography in the High Middle Ages

Albert Van Helden

Historians of ancient and medieval astronomy have long been familiar with a scheme of cosmic dimensions developed in late antiquity and passed on to Christian Europe via Moslem astronomers. Recent research has revealed that this system was, in fact, developed by the great astronomer Claudius Ptolemy (ca. 150 A.D.).[1] Scholars of medieval and Renaissance literature have, however, ignored this system of cosmic dimensions, and have even professed their ignorance of any measure of the medieval cosmos. Yet, an examination of European literature of the High Middle Ages shows that these cosmic dimensions can be found in all types of literature, from the most technical and abstruse to the most popular. It is the purpose of this essay to illustrate how widespread this scheme of cosmic dimensions actually was.

I

In *The Discarded Image* (1964), C. S. Lewis set as his task to describe for the neophyte in medieval and Renaissance literature "the imagined universe which is usually presupposed in medieval literature and art."[2]

Medieval man was, according to Lewis, "an organiser, codifier, a builder
of systems," who wanted "a place for everything and everything in the
right place."[3] In spite of the fact that his intellectual inheritance from
the Greeks and Romans was a quiltwork of often contradictory pieces,
medieval man managed to construct from them a "single, complex, har-
monious mental Model of the Universe."[4] Knowledge of this Model,
Lewis claimed, is essential to our understanding of medieval literature:[5]

> I hope to persuade the reader not only that this Model of the Uni-
> verse is a supreme medieval work of art but that it is in a sense the
> central work, that in which most particular works were imbedded,
> to which they constantly referred, from which they drew a great
> deal of strength.

Lewis went on to describe in detail this medieval cosmos—basically
the cosmos according to Aristotle—with its central earth surrounded by
the spherical shells of water, air, and fire, and then the nine spherical
shells (simply called spheres) carrying the heavenly bodies: moon, Mercury, Venus, sun, Mars, Jupiter, Saturn, the fixed stars, and the *primum
mobile*. Surrounding it all was the Empyrean heaven, the abode of God.
In such a detailed description of this medieval cosmos, the question as
to its dimensions naturally arises. How big was this Model? How far was
Jupiter away from the earth? How large were the fixed stars? Here Lewis
was not quite so well informed. He wrote:[6]

> The dimensions of the medieval universe are not even now, so gen-
> erally realised as its structure; . . . The reader of this book will al-
> ready know that Earth was, by cosmic standards, a point—it had
> no appreciable magnitude. The stars, as [Cicero's] *Somnium Scipi-
> onis* had taught, were larger than it. Isidore in the sixth century,
> knows that the Sun is larger, and the Moon smaller than the Earth
> (*Etymologies*, III, xlvii–xlviii), Maimonides in the twelfth main-
> tains that every star is ninety times as big, Roger Bacon in the thir-
> teenth simply that the least star is 'bigger' than she.

Had Lewis examined Roger Bacon's works rather than reading the few
brief statements on cosmic dimensions in A. O. Lovejoy's *The Great
Chain of Being*,[7] he would have found a complete scheme of sizes and
distances laid out in great detail. But Lewis stressed in his account early
works that were seminal in the formation of the medieval world view,
from Cicero's *Dream of Scipio* (part of his *Republic*), written in the first
century B.C., to Boethius' *Consolation of Philosophy*, written in the
sixth century A.D.[8] In this tradition we find astronomical information
such as that the earth is as a point compared to the heavens—one of the
necessary postulates of Greek spherical astronomy. But we do not find

the technical astronomy of late antiquity represented. This part of the Greek legacy reached Europe only in the twelfth century by way of the Moslem world. What, then, was the origin of the system of cosmic dimensions of which Lewis was ignorant?

II

Greek efforts to determine the size of the earth and the sizes and distances of heavenly bodies scientifically began at about the time of Alexander's conquests. In *De Caelo* Aristotle (Alexander's tutor) informs us that mathematicians had estimated the circumference of the earth to be about 400,000 stades, that is, something like 40,000 miles.[9] A generation or so later, Aristarchus of Samos (fl. ca. 285 B.C.) gave a mathematical demonstration of how one might determine the relative sizes and distances of the sun and moon. His figure that the sun's distance from the earth is more than eighteen but less than twenty times the moon's distance remained authoritative until the seventeenth century.[10] Eratosthenes of Cyrene (fl. ca. 225 B.C.) showed how, by means of a gnomon, or shadow stick, and a few simple measurements, one could determine the earth's circumference.[11] Hipparchus of Nicaea (fl. ca. 135 B.C.) estimated the apparent sizes of the planets and found the distance of the moon by means of a geometrical method that involved eclipses.[12] All these strands were pulled together by Claudius Ptolemy (fl. ca. 150 A.D.), who produced a complete system of sizes and distances.

Ptolemy's great *Mathematical Syntaxis*, better known under its Arabic title *Almagest*, is the final product of the confluence of the Mesopotamian and Greek astronomical traditions. Ptolemy found models that could successfully represent the motions of the sun, moon, and planets. From certain carefully chosen observations he determined the distance to the moon, and, using Hipparchus' eclipse diagram, he then found the distance and size of the sun as well.[13] This is as far as he went on the subject of sizes and distances in the *Almagest*. In a minor work on cosmology entitled *Planetary Hypotheses*, Ptolemy addressed himself to the question of the dimensions of the cosmos and the bodies in it. Since, however, this particular section of the book was missing at an early date from the original Greek version of the work and has survived only in some Arabic and Hebrew codices, it was not until very recently that historians discovered that Ptolemy was actually the author of the medieval system of cosmic dimensions.[14]

Ptolemy started with the Aristotelian notion of plenitude: since there were no empty spaces in the cosmos, the spherical shell containing a heavenly body had to fit snugly between the shells of the heavenly bodies above and below it. The thickness of a sphere was only just great enough to accommodate the planet's varying distances from the center and no greater, for Nature does nothing in vain. Therefore, the greatest geocentric distance of a heavenly body had to equal the least distance of the next higher body; for example, Jupiter's greatest distance was equal to Saturn's least distance. For any particular heavenly body, the ratio of its greatest to least geocentric distances could easily be found from the mathematical model in the *Almagest*. Thus, given one absolute distance—Ptolemy used the absolute distance of the moon—all other distances could be calculated. He now knew all absolute distances in the cosmos in terms of earth-radii (hereafter e.r.), and, knowing the radius of the earth, he could actually express all cosmic distances, including the radius of the cosmos itself, in stades, the common earthly distance measure.[15]

From his predecessor Hipparchus, Ptolemy adopted estimates of the apparent sizes of the sun, moon, planets, and fixed stars, and, knowing their distances, he could find their actual sizes compared to the earth's.[16] The result was, therefore, a complete set of sizes and distances (see table). The position and size of every body in the cosmos was now specified with a great show of precision (often to the nearest cubit!), even if, by our standards, the figures were in error by orders of magnitude.[17]

As usually happens in science, Ptolemy's system of sizes and distances contained some minor inconsistencies. His successors, beginning with the Moslem astronomers of the ninth and tenth centuries who were often unaware that Ptolemy was the originator of this system, tried in various ways to eliminate these imperfections. Variant schemes of sizes and distances that departed from Ptolemy's in only minor aspects were the result of these efforts. The most influential of these—more influential in medieval Europe than the original scheme of Ptolemy—was that of Al-Farghānī (ca. 800–870 A.D.), known in the Latin West as Al-fraganus. As can be seen in the table, his system varied only slightly from Ptolemy's.[18]

The *Almagest* is, even today, a forbidding treatise, clearly not intended for beginners. Al-Farghānī's *Jawāmiᶜ*, or "Summary," of the *Almagest*, which contained his version of the Ptolemaic cosmic dimensions, was designed to help initiate the beginner into Ptolemaic astonomy. It was, therefore, one of the first astronomy books translated from Arabic into Latin by Christian scholars in the twelfth century, and until it was replaced by indigenously prepared textbooks more suited to instruction in the *quadrivium* (the mathematical course of instruction in the undergraduate curriculum) at European universities, it remained the most popular introduction to astronomy in the Latin West.[19] Al-Farghānī's version of the Ptolemaic scheme of cosmic dimensions, introduced into the

The Ptolemaic System of Sizes and Distances
According to Ptolemy (P) and Al-Farghānī (F)

Body	Absolute Distance in E.R.			Apparent Diameter at Mean Distance Sun = 1	Actual Diameter Earth = 1	Volume Earth = 1
	Least	Greatest	Mean			
MOON P	33	64	48	1⅓	¼ + 1/20	1/40
F	33½ + 1/20	64⅙	48⅚	1⅓	5/17	1/39
MERCURY P	64	166	115	1/15	1/27	1/19683
F	64⅙	167	115½	1/15	1/28	1/22000
VENUS P	166	1,079	622	1/10	¼ + 1/20	1/44
F	167	1,120	643½	1/10	3/10	1/37
SUN P	1,160	1,260	1,210	1	5½	166⅓
F	1,120	1,220	1,170	1	5½	166
MARS P	1,260	8,820	5,040	1/20	1⅐	1½
F	1,220	8,876	5,048	1/20	1⅙	1½ + ⅛
JUPITER P	8,820	14,189	11,503	1/12	4⅓ + 1/40	82½ + ¼ + 1/20
F	8,876	14,405	11,640	1/12	4½ + 1/16	95
SATURN P	14,189	19,865	17,026	1/18	4¼ + 1/20	79½
F	14,405	20,110	17,258	1/18	4½	91
FIXED STARS P	—	—	20,000	1/20	4½ + 1/20	94⅙ + ⅛
F	1st Magnitude		20,110	1/20	4½ + ¼	107
	2nd Magnitude		"	—	—	90
	3rd Magnitude		"	—	—	72
	4th Magnitude		"	—	—	54
	5th Magnitude		"	—	—	36
	6th Magnitude		"	—	—	18

Ptolemy converted all distances to stades, using the value 1 e.r. = 28,667 stades. Al-Farghānī converted all distances, using the value 1 e.r. = 3,250 miles, where 1 mile = 4,000 "black cubits."

Latin educational system in the twelfth century and retained by many Latin astronomy textbooks that replaced his book,[20] thus became the common heritage of literate Europeans. It was the quantitative dimension of C. S. Lewis' *Discarded Image,* the medieval model of the cosmos.

Yet Lewis and his colleagues (for instance, A. O. Lovejoy and E. M. W. Tillyard) who have guided beginners into the difficult world of medieval and Renaissance literature have managed to miss this dimension almost completely. And if they have sometimes reported a number expressing the size or the distance of some heavenly body, they have not realized that such numbers were part of a complete system of cosmic dimensions.[21] The result has been that outside a very restricted circle of scholars of technical ancient and medieval astronomy there was virtually no realization that a coherent system of cosmic dimensions in fact existed in the high Middle Ages, that it was the common heritage of all university-trained Europeans, that it was a central part of the medieval world picture, and that it can be found across the entire spectrum of the literature of the period.

III

A brief glance at medieval literature will illustrate the centrality and universality of these cosmic dimensions. We find them in technical treatises and elementary texts in astronomy, in ponderous Latin philosophical treatises and vernacular instruments of popular edification, in sophisticated cosmological poetry and the lives of saints.

In the astronomical literature itself, the scheme is, of course, found in surviving copies of Al-Farghānī's *Jawāmi*ᶜ, twice translated and known under several different titles.[22] The most popular European textbook in astronomy, Johannes de Sacrobosco's *De Sphaera*, written around 1250, was so elementary that cosmic dimensions were beyond its scope. However, the text of this little book was almost invariably accompanied by a commentary, and these commentaries often gave the system of cosmic dimensions of Al-Farghānī.[23] The readings for the astronomical part of the *quadrivium* also included one of the versions of the *Theorica Planetarum*, textbooks in which the actual mathematical models of the seven heavenly bodies were given. The *Theorica Planetarum* of Campanus of Novara, composed around 1260, incorporated the cosmic dimensions of Al-Farghānī with some minor modifications. Campanus actually gave the dimensions of each planetary model (radius of excentric circle, radius of epicycle, amount of excentricity) in miles![24] If instead the earlier and more popular version, the *Theorica Planetarum*, falsely ascribed to Gerard of Cremona (twelfth century),[25] was used, which did not contain the cosmic dimensions, then a small tract by the Moslem astronomer Thābit ibn Qurra, entitled *Liber de Quantitatibus*, containing the scheme of sizes and distances given by Al-Farghānī, was included in the readings.[26]

In the third quarter of the thirteenth century, a certain Leopold of Austria, about whom nothing else is known, wrote a compendium of astronomy, *Compilatio de Astrorum Scientia.* This work was translated into French early in the fourteenth century, and, as one of the earliest vernacular textbooks in astronomy, it enjoyed great popularity for the next several centuries. Leopold did not ignore the sizes and distances of heavenly bodies. He wrote:[27]

> Les grandeces des cors celestres sont 14. Li Solaus a le quantitet de le Terre 100 fois 60 fois 6 fois et le 4ᵗ d'une 8ᵉ, les plus grandes estoilles en ont 115 fois, Jupiter 95 fois, Saturne 91 fois, chelles qui sont de le seconde quantité fixes estoilles 90 fois, chelles qui sont de le tierche quantitet 70 fois, de le quarte 50 fois, de le quinte 36 fois, de le sixte 18 fois, Mars une fois et le moitiet d'une 8ᵉ par une fois, Venus au rewart de le Terre le porcion de 37, li Lune 39 et un poi plus, Mercure une part de 22,000 partie de le Terre.

Leopold went on to give the distances of each of these bodies using the measures of Ptolemy rather than those of Al-Farghānī, from whom he had taken the above sizes.[28] Not surprisingly, then, the scheme of cosmic dimensions that originated with Ptolemy is commonly found in medieval astronomical literature.

It is also found in the more general philosophical literature of the period. The most celebrated instance is to be found in the *Opus Maius* of Roger Bacon, composed in 1266. After giving the diameter of the earth according to Al-Farghānī, 6,500 miles, Bacon wrote:[29]

> Alfraganus is of the opinion, from a comparison of the semidiameter of the Earth with that of the starry sphere, that the distance of the starry sphere from the center of the Earth is 20,110 times the semidiameter of the Earth, 65,357,500 miles, which if doubled will give the diameter of the whole starry sphere as 130,715,000 miles. When this is multiplied into three and one seventh we shall have the circumference of a great circle in the starry sphere, namely, 410,818,517 miles, and three sevenths of a mile, that is, 1714 cubits, and two sevenths of a cubit.

Bacon next gave the reader the length of a degree on a great circle of the sphere of the fixed stars and also the surface area of this entire sphere. He then turned his attention to the distances of the heavenly bodies according to Al-Farghānī. The following lengthy passage will illustrate his fascination with these numbers:[30]

> Moreover, the semidiameter of the starry heavens is the longer distance of the sphere of Saturn, because they join without intermediate space. But its nearest distance to the Earth is 46,816,250 miles, which is the longer distance of the sphere of Jupiter, whose nearer

distance is 28,847,000 miles, which is the longer distance of the
sphere of Mars, whose nearer distance is 3,965,000 miles, which is
the longer distance of the sphere of the Sun, whose nearer distance
is 3,640,000, which is the longer distance of the sphere of Venus,
whose nearer distance is 542,750, which is the longer distance of
Mercury, whose nearer distance is 208,541 and two thirds of a
mile, and this is the longer distance of the Moon, and this, as Al-
fraganus says, is sixty-four and a sixth of one time equal to the half
of the diameter of the Earth, and the nearer distance of the Moon is
109,937 and a half of a mile, that is, 2000 cubits, and this distance
is thirty-three times and a half of a tenth, that is, one-twentieth of
one time equal to half of the diameter of the Earth. The diameters
of the separate spheres are obtained by doubling the semidiame-
ters; the circumference of each is found by tripling the diameter
with the addition of a seventh part, and the whole surface of each
sphere is found by multiplying its diameter into its circumference,
as was explained in the case of the Earth and the starry sphere.
Any one can find these dimensions by computation, and for this
reason I omit them to avoid prolixity.

Bacon now gave the thicknesses of all the spheres in miles and calculated how long it would take a man walking twenty miles per day to
reach the sphere of the moon.[31] A few pages later he gave the diameters
and volumes of the planets and fixed stars according to Al-Farghānī, ending with the hierarchy of bodies in the cosmos according to size:[32]

From all these facts, then, that have been mentioned in regard to
the magnitudes of the heavenly bodies it is evident that greater
than all, with the exception of all spheres other than that of the
Earth, is the Sun; then in second place are stars of the first magnitude; in the third place, Jupiter; fourth, Saturn; fifth, all the remaining fixed stars according to their grades and orders; sixth,
Mars; seventh, fixed stars known to sight; eighth, the Earth, ninth,
Venus, tenth, the Moon; eleventh, Mercury.

By fixed stars known to sight Bacon meant "other stars in infinite
number, the size of which cannot be ascertained by instruments, and yet
they are known by sight, and therefore have sensible size with respect to
the heavens." The context here was the demonstration that "the Earth
does not possess any sensible size with respect to the heavens."[33]

From this demonstration of the presence of the Ptolemaic system of
cosmic dimensions, usually in the form given by Al-Farghānī, it is fair to
say that after about 1250 all educated persons in Europe were familiar
with the nesting sphere principle and the cosmic dimensions derived
from it. But if this system of sizes and distances became a common-

place, a part of the deep-seated conception of the cosmos, we might expect to find traces of it in the popular literature of the time as well. This is indeed the case.

As early as 1245, in a French vernacular poem entitled *Image du Monde*, a certain Gossouin of Metz made liberal use of Al-Farghānī's cosmic dimensions. The work, in a slightly later prose version, attained immense popularity that continued unabated well into the sixteenth century. It was translated into English and printed by William Caxton as early as 1480, under the title *Mirrour of the World*, and went through numerous editions. A popular encyclopedia, *Mirrour of the World* contained a liberal measure of astronomical information, designed to drive home the moral lesson of the vastness of God's creation. In the chapter on "How the Mone and the Sonne haue eche of them their propre heyght," we find the following information: [34]

> The earth is "more grete than the body of the mone . . . by xxix [read xxxix] tymes and a lytil more."
> The moon is "in heyght aboue the erthe xxiiii tymes and a half as moche as therthe hath of thycknes." In other words, the moon's mean distance is 49 e.r.
> The sun is "gretter than al therthe wythout ony defaulte by an C.lxvi. tymes, and thre partyes of the xx parte of therthe," that is, 166³⁄₂₀, by volume.
> The mean distance of the sun is "ffyve hondred lxxx and v tymes as moche as therthe may haue of gretenes and thyckens thurgh," that is, 1170 e.r.

In the next chapter Gossouin gave the distance of the fixed stars—Al-Farghānī's 20,110 e.r.—and brought that measure down to the experience of everyday life by translating it into day-trips: [35]

> Fro therthe vnto the heuen, wherin the sterres ben sette, is as moche grete espace; ffor it is ten thousand and .lv. sythes as moche, and more, as is alle therthe of thycknes [10,055 e.d. = 20,110 e.r.]. And who that coude acompte after the nombre and fourme, he myght knowe how many ynches it is of the honde of a man, and how many feet, how many myles, and how many Journeyes it is from hens to the firmament or heuen. Ffor it is as moche way vnto the heuen as yf a man myght goo the right way without lettyng, and that he myght goo euery day xxv myles of Fraunce, . . . and that he taried not on the waye, yet shold he goo the tyme of seven .M.i.C, and .lvii. yere and a half [that is, 7,157½ years] er he had goon somoche way as fro hens vnto the heuen where the sterres ben inne.

In fact, had Adam started on such a journey the day he was created, he would still have 713 years to go when this tract was written, in 1245.[36] Gossouin had earlier given the length of the year as 365¼ days, and the diameter of the earth as 6,500 miles.[37] The calculation of the length of Adam's journey is correct to within the accuracy of computation, and the implied date of the Creation is 5199½ B.C., the accepted date for that event.[38]

Even in popular literature as far removed from astronomy as the legends of the saints, we find traces of the cosmic dimensions. In the *South English Legendary*, composed in the vernacular in the thirteenth century, we learn in the legend of St. Michael that the earth is smaller than the smallest star,[39] and then,[40]

Biside þe eorþe in þe on half.   Þe sonne sent out hure liȝt
An hondred siþe & fiue and sixti.   as it is iwrite
Þe sonne is more þanne þe eorþe.   woso wolde iwite
And þe eorþe is more þanne þe mone.   nye siþe iwis
Þe mone þinnþþe more.   for he so nei us is
Þe sonne is herre þan þe mone.   more þanne suche þreo
Þenne it beo heune to þe mone.   þe lasse he is to seo
Muche is bitwene heuene & eorþe.   for þe man þat miȝte go
Euerich dai forti mile.   euene upriȝt and eke mo
He ne ssolde to þe heiost heuene.   þat þe aldai iseoþ
Come in eiȝte þousond ȝer.   þere as þe sterren beoþ
And þei Adam oure ferste fader.   hadde bigonne anon
Þo he was ferst imade.   toward heuene to gon
And hadde euerich day forti mile.   euene upriȝt igo
He hadde noȝt ȝute to heuene icome.   bi a þousond ȝer & mo

The writer should have used day-trips of 20 miles, not 40 miles, but the confusion is understandable since the length of the mile and league varied enormously from place to place. Al-Farghānī used a mile of 4,000 cubits,[41] but in the same demonstration Moses Maimonides (1135–1204) used a mile half as long as Al-Farghānī's:[42]

Accordingly, it has been demonstrated that the distance between the center of the Earth and the highest part of the sphere of Saturn is one that could be covered in approximately eight thousand and seven hundred years of three hundred and sixty-five days each, if each day a distance is covered of forty of our legal miles, of which each has two thousand of the cubits used for working purposes.

Roger Bacon assumed that Al-Farghānī's cubit was a foot and a half long,[43] and in translating *Image du Monde* in 1480, William Caxton assumed that xxv myles of Fraunce . . . is 1 englissh myle."[44]

Finally, we find Al-Farghānī's version of the Ptolemaic system of sizes and distances reflected in the works of Dante. In his *Convivio* we find the information that the sun is 5½ times as large in diameter as the earth, that the diameter of Mercury is only 1/28 the earth's, and that Venus' least distance from the earth is 167 e.r.[45] Dante ascribes this information to Al-Farghānī. In the *Commedia* we find the statement that Venus is the last (or highest) sphere tainted by earthly inclinations because the shadow cast by the earth comes to a point within this spherical shell.[46] The length of the shadow cone can be calculated from the eclipse diagram (see n. 12). Ptolemy and Al-Farghānī both gave it as 268 e.r.[47] Its apex was therefore in the sphere of Venus.

IV

As shown by this sample of medieval literature, the system of cosmic dimensions devised by Ptolemy and passed on to the Latin West with minor modifications by Moslem astronomers can be found on all levels of medieval European literature. By the thirteenth century it had become a regular part of the *quadrivium*, usually in the version of Al-Farghānī, and it was therefore learned by all university students. Moreover, as illustrated by the *South English Legendary* and Gossouin's *Image du Monde*, it also frequently found its way, in full or in part, into the experience of the lay public. It is puzzling, therefore, that modern writers on medieval literature who have stressed the centrality of the medieval world model to our understanding of that period's literature have ignored the quantitative dimensions of this model. Perhaps this is an example of C. P. Snow's division between the "two cultures."[48]

An Iconography of Noses
Directions in the History of a
Physical Stereotype

Alfred David

Toi, du charme physique et vainqueur,
prête-m'en: Et faisons à nous deux un hèros de roman.
  —*Cyrano de Bergerac*

Cyrano's remark to Christian that the two of them will combine to
make the hero of a novel recognizes that romantic fiction is based on
stereotypes. Ironically Christian's conventional good looks, his "charme
physique," is inhibited by a timid sense of his own mediocrity. Cyrano's
flamboyance, vitality, and indomitable spirit, on the other hand, are at
one and the same time displayed in and vitiated by his grotesque nose. It
is this feature that links him with the anti-type of the romantic hero,
the type of the boastful and vulgar peasant represented in English literature by the Miller among Chaucer's pilgrims, that jangler who tells the
most famous bawdy tale in English.

   The Miller's nose in that portrait is not only a touchstone of Chaucer's
art but, as I shall argue, a turning point in the art of literary portraiture:

> *Upon the cop right of his nose he hadde*
> *A wert, and theron stood a tuft of heeris. (I.554—55)*[1]

The precision of visual detail along with the movement and thrust of these lines convey the narrator's, and consequently also the reader's, surprise and delight in the wonders of this brave new world. The enjambment and arresting caesura after "A wert" literally make one catch one's breath before proceeding to the crowning touch: "and thereon stood a tuft of heeris." The *Middle English Dictionary* defines "cop" in this context as the "tip" of the nose, though since Skeat it has usually been glossed as "top," "summit," or "ridge." For esthetic, as well as for philological reasons, I would like to think the *MED* is correct in placing the wart at the very end.

Such a nose raises questions other than philological. There is a question, for instance, whether such striking physical details in the Prologue ought to be taken as realistic observations or symbolic expression. Is this an individualized or a typical nose? If symbolic, just what does it symbolize? Is it perhaps, in conjunction with other details, an icon of Discordia, as Professor D. W. Robertson has argued?[2] Although questions of style are often put in such terms, these are modern formalist categories, very much oversimplified, and they can be more of a hindrance than a help. We need to place the question of style in the context of continuous historical change through which esthetic modes and generic structures are undergoing transformations as they are adapted for new audiences to meet new social conditions.

The subject is vast, but in this paper I want to take it by the nose, so to speak, and look at noses as one of the indicators of change in the literary portrait, or "descriptio" as it is called in the medieval rhetoric books. Thinking historically about noses raises another question: where does the Miller's nose end, that is to say, in what directions does it point, not just in the *Canterbury Tales* but in European literature and art after Chaucer? The reappearance of the Miller's nose in the Reeve's Tale strikes me as a significant development in the history of style. Because Robin the Miller holds his nose too high, the nose of his surrogate, the arrogant Symkyn, gets broken. In the stereotyped romance portraits, as we shall see, a nose is a nose—its effect is confined to its proper place, between the forehead and the mouth. But in the *Canterbury Tales* details of physical descriptions begin to have narrative consequences.

The Miller's nose, then, points to a fascination that this central feature of our faces holds for our culture, a fascination reflected in our language, literature, art, and history. We encounter similarly arresting images in other faces among the pilgrims—the Friar's twinkling eyes and the Pardoner's glaring ones, the Monk's glistening bald forehead and face, the Summoner's pimples and boils. I could have chosen some other feature—eyes or beards, for example, are more numerous in the Prologue than noses, of which there are only two: the Miller's and the Prioress's. However, it is precisely the structural relationship between this pair of noses that illustrates my thesis that representational form in art is a function of historic process. To see what Chaucer has done with the Prioress's nose and the Miller's, we shall take a closer look at some earlier literary noses and then make a survey of some noses that follow after.

Noses are a common though by no means inevitable item in the formal portraits that begin to crop up in the twelfth-century *romans d'antiquité* like the *Roman de Thèbes* or the *Roman de Troie* and in the romances of Chrétien. Benoît de Sainte Maure, for example, introduces a whole section of portraits into the *Roman de Troie*, starting with the Greeks—Castor and Pollux, Helen, Agamemnon, Achilles, and so forth—followed by the Trojans. These run from as little as four lines (Menelaus) to sixty-eight lines (Hector). One of the lengthier portraits is that of Troilus:

> Troilus was marvellously handsome, his expression was glad, his face ruddy, his countenance bright and sincere, his forehead broad, very much becoming to a knight. He had light blond hair, very comely and naturally shiny, gray eyes full of gayety; never was anyone so handsome. When he was in a good mood, he had such a sweet look that it was a pleasure to behold him. But one thing I tell you truly that toward his enemies he had a different semblance and countenance. His nose was high and just the right size [par mesure]; his body was well-suited to arms. He had a well-made mouth and beautiful teeth, whiter than ivory or silver. (5392–5410)[3]

The portrait runs on for another thirty-six lines.

What is the purpose of such descriptions? They may owe something, though not as much as has sometimes been thought, to advice given in the medieval arts of poetry like those of Matthew of Vendôme and Geoffrey of Vinsauf, who reinterpreted classical rhetoric for the Middle Ages.[4] Both of these authors, but especially Matthew, greatly elaborate the sparing advice found in Cicero, Horace, and other classical treatises on rhetoric and poetry that the description of characters should be appropriate to age, rank, profession, nationality, and so on, and these authors provide models—a description of a pope, of Caesar, of Ulysses, of Helen—as well as references to portraits in ancient Latin authors.

These facts are well known. What has never been adequately explained is the reason behind this vogue for physical descriptions in twelfth-century romance and for the extraordinary emphasis given to the description of persons by these rhetoricians. How did Matthew of Vendôme come to regard description "comme l'objet suprême de la poésie," as Edmond Faral puts it?[5] What accounts for the phenomenon that the faces of romance heroes and heroines, including in many cases their noses, acquired a prominence that they had not held before? The author of the *Chanson de Roland* was not interested in the noses of Charlemagne and his peers. Why did a later audience want to hear about the shapes and proportions of the noses of Nestor, Troilus, and Cligés?

The idea that the physical features of our faces are an outward expression of our characters is, of course, timeless, and it is the root idea of the science of physiognomy, which was invented by the ancients. The ur-physiognomical text is a work of the third century B.C., long attributed to Aristotle. Pseudo-Aristotle bases his interpretation of physiognomy on resemblances between the human and animal kingdoms. Thus, if one looks like a lion or an eagle, then one will probably have those characteristics we know lions or eagles possess. In his chapter on noses, pseudo-Aristotle offers the following observations, which were frequently copied by later writers in Greek, Latin, and Arabic:

> Those that have thick extremities to the nostrils are lazy; witness cattle. Those that have a thickening at the end of the nose are insensitive; witness the boar. Those that have a sharp nose-tip are prone to anger; witness the dog. Those that have a circular nose-tip, but a flat one, are magnanimous; witness the lions. Those that have a thin nose-tip are bird like; but when it is somewhat hooked and rises straight from the forehead they are shameless; witness ravens; but those who have an aquiline nose with a marked separation from the forehead are magnanimous; witness the eagle.[6]

Such doctrines unquestionably influenced classical literature. Suetonius, for instance, draws portraits of the caesars that offer clues to their individual characters. Of Augustus he writes:

> His teeth were wide apart, small, and ill-kept; his hair was slightly curly and inclining to golden; his eyebrows met. His ears were of moderate size, and his nose projected a little at the top and then bent slightly inward.[7]

These details, according to Edith C. Evans, are evidence of an excellent character.[8] The nose, apparently aquiline, is a sign, as we have seen in pseudo-Aristotle, of magnanimity. In contrast is Suetonius's portrait of Caligula, whose physical appearance, according to Evans, "is especially associated with the disagreeable features of the panther and the goat." Unfortunately Suetonius does not mention Caligula's nose. Possibly it was a snub-nose that, says pseudo-Aristotle, shows that a man is lecherous; witness cocks.

These classical portraits, however, are quite different in form and purpose from those we encounter in twelfth-century literature. For one thing, the face in ancient literature primarily expresses character, not physical beauty. To a degree faces are stereotypes—thus an aquiline nose is part of the imperial look. Nevertheless, these faces are also individualized—no two are identical. What impresses one about twelfth-century faces, however, is their uniformity. Their noses are interchangeable.

Alice Colby has counted twelve beautiful noses among thirty-three portraits in twelfth-century romances—not an overwhelming number but enough to show that the nose, like the more common features such as the hair, eyes, and mouth, is one element of the stereotype.[9] These noses are variously said to be "droit," "haut," "bel," "bien seanz," "bien assis," and most frequently of all "bien fait." One need not ask, of course, why the romance heroine must be beautiful and the hero handsome, but such a catalog of clichés is surely not the only way or the most effective means to that end. Doubtless the medieval passion for amplification, urged by the rhetoricians, has something to do with it, but just what is it that is being amplified in these portraits? What the details finally add up to is that these people belong to the court and possess the qualifications ideally required of a courtly person. The qualifications include, of course, noble birth but also elegant manners and a stylized beauty. Collectively the details create a generalized type; they do not actually describe anything, but their presence is required as a symbol of value like a trademark. Thus, physical appearance becomes one badge of social class, and a nose "bien fait" lets us know, along with the rest, that all these ancient Greeks, Trojans, and Britons belong to a knightly caste that transcends the boundaries of time and nationality. The new importance attached to outward appearance suggests, like the rise of romance itself, a class newly conscious of itself and seeking self-definition. That definition is no longer based mainly on hierarchy but on signs that include items like fashionable clothing and well-shaped noses.

Class consciousness, however, involves an awareness of some opposite. "Ideal beauty" has its counterpart in what Colby terms "ideal ugliness."[10] The same elements found in beautiful portraits are grotesquely ugly in the description of the herdsman who directs the knights in Chrétien's *Yvain* to the magic spring.[11] He is black as a "moor" with a huge head, ears like an elephant, mouth like a wolf, teeth like a wild boar, a nose like a cat, that is, a flattened muzzle. Aucassin in his search for Nicolette encounters a similar apparition:

> He was tall, weird, and alarmingly ugly. He had a great mop of a head as black as smut with eyes set a palm's width apart, broad cheeks, an enormous flat nose with cavernous nostrils, thick lips redder than underdone meat and great ugly, yellow teeth.[12]

These are faces of the "vilain," the peasant, a creature made to seem as though he belonged to a different species, closer to the animal world, the realm of nature, than to the world of the court. Stylized as they are, these descriptions nevertheless suggest that the aristocratic audience held a stereotyped view of the peasantry just as it did of itself. For the modern reader these portraits of "ideal ugliness" are more interesting than their courtly countertypes. These wild men preserve something of the green world that the elegant knights have lost. The author of *Gawain and the Green Knight* certainly sees that, and perhaps Chrétien does too. However that may be, these portraits set up a polarity, implicit even when we see only portraits of ideal beauty, between "vileinye" and "curteisye."

Such a polarity can be seen even more clearly in an allegory like the *Roman de la Rose*. Though Guillaume de Lorris does make some attempt to differentiate the personified dancers in the Garden of Mirth, the overwhelming feeling one gets from the lengthy descriptions of their faces and bodies is of sameness. One could not say whose nose is "of good proporcioun," "by mesure wrought ful right," "wel wrought," "wrought at poynt devis," or so "fair" as "hath no womman alyve."[13] The dancers, as is often observed, are opposites of the portraits painted on the outside of the wall around the garden: Mirth and Gladness correspond to Sorrow, Youth to Elde, Largesse to Coveitise. Yet the portraits on the wall are actually much more varied than those in the garden and belong to a different stylistic tradition—that of moral allegory, such as descriptions of the deadly sins. There are fewer physical details and these usually convey a direct allegorical meaning. Thus Coveitise has crooked hands; Envy squints with one eye shut; Elde wears a fur cloak because old people are always cold. The only nose in the set belongs to Hate, which, in the Middle English translation, is "snorted up for tene" (157). Villainy, which immediately follows Hate, is given the sketchiest of all the portraits, amounting to no more than three or four lines of the most generalized description.

There is, however, a more detailed portrait of Villainy within the garden that answers to the descriptions of the ugly herdsman in *Yvain* and the ugly peasant in *Aucassin*. This is Daunger, who appears as follows in the Middle English translation:[14]

> *Full gret he was and blak of hewe,*
> *Sturdy and hidous, whoso hym knewe;*
> *Like sharp urchouns his her was growe,*
> *His eyes reed sparclyng as the fyr glowe,*
> *His nose frounced, full kirked stood. (3133–37)*

The black complexion, the bristly hair, the nose wrinkled and crooked, all belong to the stereotype, and Daunger is repeatedly called a "vilain." When Daunger is caught sleeping on the job of guarding the rose, Shame and Dread, who personify qualities of the lady, rouse him and chide him for acting out of character: he should be "dogged," "froward," and "outrageous," like the churl that he is.[15] Plate 1, from a fourteenth-century manuscript, shows how the artist has differentiated the noses of the ladies from the peasant nose of Daunger. He has also suggestively emphasized Daunger's "burdoun," the great club that is the "vilain's" weapon.

The club, like the whole description of Daunger, brings out a meaning latent in most of these romance portraits. The symbols of class also have a sexual valence. Daunger starts up out of a hiding place at the moment the Lover asks Fair Welcome to let him pluck the rose. Although Daunger is said to be an attribute of the lady, we probably restrict the meaning of the allegory by reading into it too novelistic a plot. In a sense *all* the personifications in the garden are projections of the lover's psyche, just as it is his own image he sees in the well of Narcissus. Thus the churl Daunger, who brandishes his great burdoun as the herdsman in *Yvain* carries a huge club and the peasant in *Aucassin* a quarter staff, may be seen as an image of the gross desire that is constantly coming between the courtly and later the Petrarchan lover and the object of his desire, a desire that he must learn to suppress. Thus, the opposition between "curteisye" and "vileinye" in romance is both social and sexual in nature. On the one hand there is base lust; on the other, "fine amour"; and these differences in spirit are reflected in the face, costume, language, and behavior.

The nose has, of course, a special relevance for such a reading because its shape and location make it a natural phallic symbol, although our notion of "symbol" hardly does justice to the symbiotic connection that was believed to exist between the nose and the sexual organs. That relationship is spelled out clearly in the most famous of all Renaissance physiognomy books, *De humana physiognomónia* by Giovan Battista della Porta, the first of the many editions of which appeared in 1586. In the chapter "Del Naso," Porta states that "Il naso risponde alla verga" (the nose corresponds to the "rod," that is, the penis). Whoever has the one long and big, or pointed and big, or short, the same may be judged of the other; likewise the nostrils correspond to the testicles. Men who are especially virile are called "nasuti" (big noses). The Emperor Heliogabalus, Porta continues, who practiced "soverchie delizie nefanda libidine" (sovereign delights of nefarious lust), gathered "un esercito di uomini stalloni" (an army of studs), which he personally selected from such "nasuti" to serve his infamous lust.[16] Walter Curry long ago argued that

Plate 1. Daunger, Shame, and Dread. From MS. 48 of the *Romance of the Rose* in the Pierpont Morgan Library.

Chaucer's portraits of the Miller and Reeve, among others, were based on physiognomy books.[17] In connection with the Miller's nose he quotes one compendium of such lore: "Now let us treat of the Nose, which . . . relates to the Genitals or Secrets. When a Mole is on the root of the Fore-head, there is another on the Foreskin of the flesh."[18] Although such passages testify to the sexual significance of noses, they do not necessarily justify Curry's conclusion that in the portrait of the Miller "Chaucer has made ample use of the science of physiognomy."[19] The fact is that the "science" of physiognomy is no more than a learned codification of popular beliefs and superstitions that have always found expression in art and literature. The Wife of Bath requires no great learning to know that there is a connection between the mark of Mars upon her face and the mark that she has "in another privee place" (III.619–20). Physiognomy could only have confirmed what Chaucer and his audience understood about noses, warts, and facial hair. The modern reader also instinctively understands the Miller's nose, even if he has never heard of physiognomy. The whole portrait of the Miller, from his door-butting pastimes to his bagpiping, is loaded with sexual imagery.

That same imagery is present in the Reeve's Tale, where Oswold makes the Miller pay through the nose in a double sense. The clerk "swyve" Symkyn's wife and daughter, and in the fight afterwards they smash his "camus" (flat) nose. That nose is a badge of the "auncestrye" of which this miller is inordinately proud, and part of the endowment he has transmitted to his daughter:

> The wenche thikke and wel ygrowen was
> With camus nose and eyen gray as glas. (I.3973–74)

The last line is an epitome of what happens to the romance stereotype in the *Canterbury Tales*. In it ideal beauty and ugliness are conflated amusingly in a single description—the peasant nose with the noble eyes. The model for such a line is, of course, the line describing the Lady Prioress: "Hir nose tretys, hir eyen greye as glas" (I.152).

Exactly that kind of comic relationship exists between the noses of the Prioress and the Miller in the General Prologue. When we isolate such details from their context and examine them in the perspective of romance literature, we can see in them the vestige of a deep structure common to romance, fairy tale, and moral allegory: a binary opposition between "curteisye" and "vileinye," good and evil, virtue and vice, Beauty and the Beast. The purpose of making such an observation, however, is not to reduce Chaucer's Prologue to an empty schema but to show how completely that schema has been transformed and revitalized as the old value structure itself has been changing.

In romance and moral allegory external details establish a clear structure of values, whether those values are social or ethical and religious. The old value structure survives, but set now against a secularized, multi-layered, commercial, and increasingly urban society, like that of fourteenth-century London, it appears anachronistic or, in Donald Howard's term, "obsolescent."[20] The genuinely ideal portraits in the Prologue, that is ideal in the moral sense, contain very little physical description. The morally ideal pilgrims—the Knight, the Parson, and the Plowman—are described primarily in terms of what they have done or what they do (or will *not* do). All we actually see of the Knight is his rust-stained *gypoun*, a detail more truly expressive of inner worth than shining armor would be; of the Plowman we see only his humble tabard and the mare he rides; the Parson we do not see at all as he appears on the pilgrimage, but only in the image of the good shepherd, "Upon his feet, and in his hand a staf" (I.495). In contrast, the other pilgrims are created out of a world of appearances and possessions, objects worn ostentatiously like the Merchant's beaver hat or objects of desire like the Clerk's "twenty bookes clad in blak or reed" (I.294). We are still looking at portraits of types of people, scanning their faces for signs of value. But the types are no longer simple stereotypes, the details do not sort themselves into easily grasped patterns, and the values have grown elastic; that is, they can be stretched to apply to many different things.

Thus the "tretys" nose of the Prioress retains its familiar class associations, but its significance is no longer static but qualified in juxtaposition—one could say collision—with other details in the portrait of the Prioress herself and in other portraits. "Ful wel she song the service divyne, Entuned in hir nose ful semely" (I.122–23). Of course, it is the holy office that is "entuned" ful semely in her nose, or is it by chance "entuned" in her ful semely nose? It does not really matter for both the nose and the religious service partake of a "seemlinesse" that characterizes a variety of objects and activities associated with the Prioress—table manners, a pleated wimple, a piece of jewelry. One can say that in this society style becomes a negotiable commodity to be inherited, worn, bought, learned, or acquired like any other possession. The Miller's virility is another sort of commodity more in demand at this time than the seemliness of the Prioress or the courtesy of the Knight. His nose signals a vital presence beside which the Prioress's nose seems recessive. The fashion in noses is changing.

In the fourteenth century kings are still pictured in accordance with an idealized type. Thus, in the Wilton diptych (Plates 2 and 3) Richard II is portrayed as a youth, though he was in his thirties when the painting was made.[21] His nose is "bien fait," and the artist has given almost identical noses to the Virgin and the angels, particularly the angel kneeling at the Virgin's feet, who is the counterpart of the king in the central panel of the diptych. The Virgin's face with its "tretys" nose portrays the ideal type that medieval nuns, including Chaucer's Madame Eglentyne, would have wished to emulate in their appearances. Like the Prioress, the Virgin has a "fair forehead." Her mouth is "ful smal." Her headdress is pleated just as the Prioress's wimple. Madame Eglentyne has often been accused of worldliness, but if she is worldly, one can see how she comes by it.

One need only leaf through John Pope-Hennessy's *The Portrait in the Renaissance* to witness a revolution in taste and style that shows nowhere more plainly than it does in the nose.[22] Among an extraordinary gallery of individualized noses, that in Jean Clouet's portrait of Francis I (Plate 4) is but one splendid example. What is idealized and immortalized in such portraits is not a type of spirituality or nobility but a physical likeness. The faces, and especially their noses, proclaim pride, authority, virility. Of such wealthy and powerful men one may say that they are truly the lords and owners of their faces.

The Renaissance brought a revival of physiognomy among other ancient arts and sciences. The nose once again became the expression of

Plate 2. The Wilton Diptych. Details, left and right panels.

Plate 3. The Wilton Diptych. Right panel.

Plate 4. Francis I of France by Jean Clouet.

character, not an emblem of ideal beauty or ideal ugliness. The art of printing added a new attraction to phyiognomy books because the text could now be illustrated by engravings. Della Porta reintroduced the device, which goes all the way back to pseudo-Aristotle, of interpreting human physiognomies by analogy with animals, and the illustrations of these resemblances no doubt contributed much to the great popularity of his work. The nose of the great humanist Angelo Poliziano (Plate 5) is discussed among others under the subdivision "naso molto grande." Della Porta says of him that his nose was "extremely disproportionate and, therefore, he had a piercing and jealous mind, praising his own works and mocking those of others."[23]

This is the literature that fascinated Walter Shandy, whose principal hobbyhorse is a translation of Hafen Slawkenbergius's monumental treatise, *De nasis*. Tristram's father attributes the rise and decline of the Shandy fortunes since the reign of the Tudors to their noses:

> He would often declare, in speaking his thoughts upon the subject, that he did not conceive how the greatest family in England could stand it out against an uninterrupted succession of six or seven short noses.—And for the contrary reason, he would generally add, That it must be one of the greatest problems in civil life, when the same number of long and jolly noses, following one another in a direct line, did not raise and hoist it up into the best vacancies in the kingdom.[24]

Walter Shandy's ruling passion sounds like the *reductio ad absurdum* of physiognomy. His *idée fixe*, however, anticipates the serious endeavors of people who would be measuring skulls, feeling cranial bumps, and scrutinizing the angles of noses. The main difference between the old physiognomy and the new lies in the latter's claim to empiricism and mathematical objectivity. Almost from its inception, however, the new science lent itself to subjective judgments based on art. Once people had learned to measure such things as facial angles, it was a short step to their declaring that certain angles are superior to others. Winckelmann is one of the first to find a perfection in Greek statues that serves as an ideal type against which to measure other physiognomies. The type of the Greek nose, he found, was inherently more beautiful than the type of the Semitic nose.[25]

The physiognomical work whose fame surpassed all others is the *Physiognomische Fragmente*, published in four huge volumes, 1775–78, by Johann Casper Lavater, a Swiss minister.[26] Lavater has sometimes been called "the father of physiognomy," though what he chiefly did was to make physiognomy fashionable and to give it the appearance of a science by documenting its "laws" with hundreds of examples. Lavater's

Plate 5. Angelo Poliziano from della Porta's *Human Physiognomy.*

book is exactly what the title says—a series of fragments—essays on a variety of physiognomical topics. The work is quite unsystematic, but it is pervaded by a central idea, namely that "the beauty and ugliness of the face have a true and precise relationship to the beauty and ugliness of the moral disposition . . . the better morally, the more beautiful; the worse morally, the uglier" (I.63). The popularity of Lavater's work is owing in large measure to a wealth of illustrations, sumptuously reproduced, of faces, silhouettes, hands, eyes, noses—many belonging to famous people—which the author analyzes in ebullient prose. The following comment on the silhouette of a head in an essay entitled "Notes toward a physiognomical treatise" may be taken as a typical sample of Lavater's style and method:

> Here follows a free-hand cutout of a universally known head. If this nose (it might be a little broader at the end and just a hair longer)—if the mere outline of this nose does not manifest intellectual curiosity and studious zeal, nor enterprising spirit, nor industry; if something by nature noble, free, and magnanimous does not reside there—that is to say, if one shows me a nose resembling this that is not distinguished by such a character . . . away with physiognomy! at least with mine! (IV.16–17)

The object of this effusion is the Emperor Joseph II, and it illustrates, among other things, Lavater's obsequious regard for the nobility and for important men in general.

Part 4 of Volume IV, "Concerning several individual parts of the human countenance," has a section entitled "A word about the nose" (257–59).

There is no such thing, Lavater says, as an ugly face with a beautiful nose: "An ugly person may have fine eyes, but not a handsome nose. I meet with thousands of beautiful eyes before one such nose, and wherever I find the latter, it denotes an extraordinary character." He goes on to list the nine requisites of a "perfectly beautiful nose." "Such a nose," he concludes, "has greater value than a kingdom." He concedes that there are excellent, even great men (for example, Socrates), with faulty noses, but these men are of a passive, suffering character. He also classifies ethnic noses:

> The Tartars, generally, have flat, curved noses; African blacks, stub noses; Jews, for the most part, hawk noses. The noses of Englishmen are seldom pointed, but generally round. If we may judge from their portraits, the Dutch seldom have handsome or very significant noses. The great men of France, in my opinion, have the character of their greatness generally in the nose. (258)

The national and racial characteristics evident in noses are a recurrent theme in Lavater's writings. Plate 6 shows a Spaniard, a Dutchman, a Moor, and "an American native of Virginia." The first two illustrate contrasting features of pride (Spain) and humility (Holland). Of the Moor Lavater notes: "The curvature of the entire face; the breadth of the eyes; the squashed nose; but especially the thick, protruding, tough lips, with the absence of all refinement and grace—are moorish characteristics." On the Virginian he observes, "How very much more noble, kind, sensitive, than the Moor! What femininity throughout! What pure vaulting of the skull. . . . Those deeply sunken eyes, are they not firmly connected with the hollowness in the profile of the nose? And does one find them in any other than loving people?" (IV.310–11) Evidently the eighteenth-century ideal of the "noble savage" has influenced Lavater's physiognomy.

One would gladly dismiss Lavater as a curiosity of the past were it not for the insidious consequences of the ideas and the mentality that he popularized and to which he gave a measure of respectability by presenting them as another advance of the enlightenment, a science purged of ancient superstition and error, handsomely printed and expensively illustrated. Lavater was certainly not an original thinker, and I do not believe that his work ever exerted any truly profound influence. Rather, he articulates and confirms prejudices already held by many of his readers. Among his contemporaries he was both greatly admired and sharply criticized. In the nineteenth century he was still remembered and cited respectfully, notably by Balzac, and of course by later physiognomists who regarded him as one of the founders of their science. In the twentieth century he has been all but forgotten, though he may be undergo-

Plate 6. National faces from Lavater's *Physiognomical Fragments.*

ing a revival. In 1969 a Swiss publisher brought out a splendid facsimile edition of the four volumes of the *Fragment.* In 1982 Princeton University Press published *Physiognomy in the European Novel* by Graeme Tytler, who regards Lavater as a major influence on nineteenth-century fiction. Tytler calls Lavater "a unique genius."[27] "What astonishes us," he writes, "is Lavater's remarkable capacity for interpretation whether he reads intelligence in the shape or position of someone's eyebrows, notes goodness in the lower part of a nose, or detects genius in a half-closed eye."[28] Though conceding that Lavater was guilty of some excesses,

Tytler nevertheless concludes that "Lavater made physiognomy attractive by his insistence on its scientific aspects. Indeed, his discussions of the bone structure, heredity, and the analogies between man and animals are among the best parts of his work and in some ways anticipate evolutionary theories."[29] The naive view in this statement, and indeed throughout the whole book, about what constitutes science may give one pause. To me, Lavater seems a good deal closer to pseudo-Aristotle than he does to Charles Darwin. Quite clearly physiognomy is not dead and bids fair to become a fresh interdisciplinary field for academic publication.

The fact that a literary scholar like Tytler seems entirely innocent of the more sinister implications in Lavater's work makes it all the more imperative to place that work into historical perspective. Lavater is still of interest to us, not, I think, because he taught novelists to read character in the face but because his writings are symptomatic of intellectual, social, and political attitudes during the pre-revolutionary, pre-romantic period that paved the way for modern racism.

Lavater himself was not a racist, but as one can see in his comments on the Moor and the Virginian, he is already thinking in racist terms, though curiously skin color does not seem to be a consideration for him. He espouses no ideology except Christianity, and he is free from the fear, suspicion, and hostility that characterize the racist writing about those not of his kind. Yet Lavater's stereotyping of faces created the climate for racism. It established polarities that the racist and xenophobe would seize on and exploit.

Lavater's concept of a "*perfectly* beautiful nose" may remind us of the "bien fait" noses of twelfth-century romance. To be sure, his nine criteria for judging noses, which include calculating the proportions of the nasal hypotenuse to its base, and of the length of the nose to the length of the forehead (3 to 1 and equal, respectively), are a good deal more complex than the simple observation in the medieval portraits that a nose is well made or that it is "par mesure," neither too big nor too small; nevertheless, the principle is the same. Value judgments of rank, intelligence, and moral character coincide with esthetic judgments, but of course such judgments are circular: that which we hold to be noble, wise, and good seems beautiful, and that which we hold to be beautiful seems noble, wise, and good.

The Lavaterian view also asserts a standard of ideal beauty that makes it possible to say that even so magnanimous a nose as an emperor's may be a hairbreadth too short. Inevitably, ideal beauty also conjures up its antithesis—ideal ugliness. The latter continues to be a class ideal. Thus, Dickens describes the Artful Dodger as a "snub-nosed, flat-browed, common-faced boy . . . with rather bowlegs and little sharp ugly eyes."[30]

But in the nineteenth century another stereotype and binary structure emerges, based not on class but on race. The snub-nosed peasant or urchin is overshadowed by the more threatening image of the hawk-nosed Jew.

It is as a protest against this recently developed physical caricature of the Jew that we can understand Sir Walter Scott's portraits of Isaac and Rebecca in *Ivanhoe*, which are more detailed with respect to physiognomy than any other descriptions of characters in the novel. They are the only characters whose noses are mentioned. Scott goes out of his way to inform his reader that Rebecca's dark, exotic, and voluptuous beauty is a variant in the taxonomy of romance, which is a match, point for point, with the conventional type:

> The figure of Rebecca might indeed have compared with the proudest beauties of England, even though it had been judged by as shrewd a connoisseur as Prince John. Her form was exquisitely symmetrical, and was shown to advantage by a sort of Eastern dress, which she wore according to the fashion of the females of her nation. Her turban of yellow silk suited well with the darkness of her complexion. The brilliancy of her eyes, the superb arch of her eyebrows, her well-formed aquiline nose, her teeth as white as pearl, and the profusion of her sable tresses, which, each arranged in its own little spiral of twisted curls, fell down upon as much of a lovely neck and bosom as a simarre of the richest Persian silk, exhibiting flowers in their natural colours embossed upon a purple ground, permitted to be visible—all these constituted a combination of loveliness, which yielded not to the most beautiful of the maidens who surrounded her.[31]

Scott's choice of the eagle instead of the hawk to describe the Jewish nose is intended to convey precisely the association of royalty and magnanimity that we have noted in pseudo-Aristotle and Suetonius. Rebecca is, as Prince John observes in this scene, "the very bride of the Canticles," a bride fit for a king.

But it is in Rebecca's father, Isaac, that we can see through the clarity of Scott's historical imagination the superimposition of two images:

> Introduced with little ceremony, and advancing with fear and hesitation, and many a bow of deep humility, a tall thin old man, who, however, had lost by the habit of stooping much of his actual height, approached the lower end of the board. His features, keen and regular, with an aquiline nose, and piercing black eyes; his high and wrinkled forehead, and long grey hair and beard, would have been considered as handsome, had they not been the marks of

a physiognomy peculiar to a race, which, during those dark ages, was alike detested by the credulous and prejudiced vulgar, and persecuted by the greedy and rapacious nobility, and who, perhaps, owing to that very hatred and persecution, had adopted a national character, in which there was much, to say the least, mean and unamiable.[32]

The exquisitely symmetrical figure of Rebecca, her brilliant eyes, her aquiline nose are inherited from her father, but in him that beauty has been distorted, not so much by age, but by prejudice and hatred, which have even reduced his stature through "the habit of stooping."

Scott called *Ivanhoe* a "romance," and it is set in the twelfth century. But, as Georg Lukács has brilliantly demonstrated, Scott's novels involve a "necessary anachronism" through which the historical realities of Scott's own age are projected backward into history.[33] In *Ivanhoe* Scott is not showing us medieval antisemitism but the emergence of modern antisemitism. Medieval antisemitism sees the Jew as a demonic figure, but he is not visualized as a racial stereotype because in the Middle Ages the concept of race, at least as we have come to know it, does not yet exist. Scott, writing after the emancipation of the Jews during the Enlightenment, *does* recognize the racial stereotype and seeks to counteract it by introducing two noble Jews into his historical romance. His vision of "those dark ages" in which the Jew "was alike detested by the credulous and prejudiced vulgar, and persecuted by the greedy and rapacious nobility" is less an image of the past than of the future. The fearful figure of Isaac steps on the stage of modern history.

For the Middle Ages the ultimate threat finds expression in the image of Behemoth, mounted upon Leviathan, portrayed in Plate 7, from a fifteenth-century Apocalypse. The artist has portrayed Behemoth as a medieval fiend with a grotesque head and with a second face sometimes seen in medieval demons, which may be influenced by legends of the sort Othello tells Desdemona about "men whose heads do grow/Beneath their shoulders" (I.iii.144–45). Behemoth has yet a third face, which is chiefly nose. Nothing could express more graphically or more grossly the latent sexuality that is feared in the dark face of the other, whether that other be the "vilain," the Jew, or the black. The phallic implications of the nose have been totally suppressed in the writings of Lavater and nineteenth-century physiognomists who followed him, but they are always present.

Plate 8 is an example of Nazi propaganda reproduced in George Mosse's book, *Toward the Final Solution*.[34] It is directed against sexual contact between Aryan and Jew, specifically, of course, male Jew and female Aryan, a crime under the Third Reich called *Rassenschande*. Here once

Plate 7. From an Apocalyse in the Musée Condé, Chantilly.

again one sees the depersonalized, "tretys" nose in profile against the
phallic nose of the monster, the romance Behemoth of the fascist imagination. What we can see beyond the end of the Miller's nose is finally
not amusing.

   "'Tis a pity," Walter Shandy muses, "that truth can only be on one
side, Brother Toby—considering what ingenuity these learned men have
all shown in their solution of noses."[35] I have tried to show that people's
solutions of noses depend on historical circumstances and that their estimates of noses are shaped by other beliefs and values inherent in the
language and culture of their time. To recognize that much is to take one
step toward understanding the true solution if there is really only one.

   Literature helps to solidify powerful stereotypes; great literature can
also be a means of dissolving those stereotypes in laughter. The first
fragment of the *Canterbury Tales* puts the "tretys" nose of romance in

Plate 8. From *Antisemitismus in Wort und Bild*, ed. Robert Korber and Theodor Pugel. Berlin, 1935.

comic perspective; it helps us not only to see through the affectations and pretensions of prioresses and millers but, if we think about it, to learn something of the roots of our prejudice. Chaucer finally had only a limited interest in noses and in physiognomy. He only gives us one other portrait of a nose in the *Canterbury Tales* in parodying the hack work of the romancer—the nose of the bourgeois knight Sir Thopas:

> *And I you tell in good certeyn,*
> *He had a semely nose. (VII.728–29)*

Stereotypes have power over us only so long as we fail to recognize them for what they are. Chaucer understood that stereotypes are a way learned men, poets among them, have of leading people around by the nose. Had Walter Shandy been able to put the question to the ghost of Geoffrey Chaucer, he would surely have asked: "Sir, what is your solution of the causes of long and short noses?" I should like to think that Chaucer would have replied in words like those of Uncle Toby, though perhaps not without irony: "There is no cause but one why one man's nose is longer than another's, but because that God pleases to have it so."

FIVE

The Influence of Alchemy
on Newton

Richard S. Westfall

The influence of alchemy on Isaac Newton was not a static condition that can be analyzed as though it happened once and for all and underwent no change. The topic demands a chronological approach. Like the tide, Newton's concern with alchemy rose and fell, and like the tide it deposited some of its burden, which remained on the shore altering its contour. We can understand the phenomenon only if we follow it through the entire cycle.

Let me torture my simile for one last analogy. We can identify what the tide has brought in only if we know what the shore was like before. Newton's interest in alchemy was not his earliest intellectual activity. It began against a well-defined background. Unless we survey the background in detail—the shore at low tide—we cannot hope to assess the changes the tide wrought. Fortunately Newton was good enough to preserve a record of the background, apparently aware that in the twentieth century a strange breed of academics would arise, a race much given to organizing symposia on sundry, arcane topics, who might find in his manuscripts food for talk and even perhaps for thought. With rare forbearance he refrained from destroying the miscellaneous papers that few men keep. As a result, we possess even the records of his undergraduate studies; I turn to them to sketch the setting for my story.

More than fifty years after his undergraduate career, in a conversation with an itinerant Venetian noble, the Abbé Antonio-Schinella Conti, who lived for a time in London and became his confidant, Newton confessed that he had been converted to Cartesianism in his youth.[1] We can in fact follow his conversion in reading notes from his undergraduate days. From all appearances, the official curriculum had nothing to do with the process. The universities were the most reactionary institutions of seventeenth-century society. By the early 1660s a revolution in thought, perhaps the most profound revolution the entire intellectual history of European civilization has witnessed, had been in progress for more than two generations. It had not yet seriously penetrated the curricula of the universities, which continued to feed the students on an Aristotelianism nearly exhausted of sustenance after four centuries of steady use. Newton's reading notes reveal that he too, in his turn, was served up with the same stale fare, readings in Aristotelian logic, rhetoric, ethics, physics, and philosophy in general as they appeared in the textbooks of the Stagirite's seventeenth-century epigoni. When I said that the recent revolution in European thought had not penetrated the walls of the universities, I meant officially. We have independent testimony that Descartes, for example, was being read unofficially in Cambridge, that his philosophy was spreading underground through what one witness called the brisker part of university, and Newton's notebook establishes that it made its way to the chamber in Trinity College that he shared with John Wickins.[2] He left the notes he was taking from various peripatetic textbooks incomplete, simply abandoned the established curriculum on which his own academic future might hinge, and plunged into an extended program of reading in the new philosophy. If we allow the evidence in his notebook to guide us, we may choose to question his assertion to Conti that he became a Cartesian. Rather, we may find it necessary to place a broad interpretation on the word "Cartesian" so that it means, not the specific philosophy of Descartes, but a somewhat less precise mechanical philosophy. Certainly Newton read Descartes. He did more than read him. He devoured him, and nearly every page of the notebook in question contains evidence of the feast. Descartes was not the only author on his menu, however. He read as well Galileo's *Dialogue*, Kenelm Digby, Joseph Glanville, Henry More, Robert Boyle, and Walter Charleton's English epitome of Pierre Gassendi's rendition of atomic philosophy. The latter especially caught Newton's attention, and the record of the notebook strongly suggests that already as an undergraduate he found the Gassendist atomic philosophy more attractive than the Cartesian. For my purposes the distinction does not matter. Let me say merely that before he took his Bachelor's degree in 1665, Newton had embraced the mechanical philosophy of nature. I take this to be what he had in mind when he told Conti that he became a Cartesian.

The mechanical philosophy of nature, largely the creation of the generation preceeding Newton, was an attempt to separate the psychic and the spiritual from the physical and material and to establish, within the realm of natural philosophy, the autonomy of material causation. Descartes's *Meditations* provided the cornerstone of its metaphysical foundation. In the sixth and last meditation, using the results of the previous five, Descartes demonstrated to his own satisfaction the existence of the material realm, which he had called into question at the very beginning of his process of systematic doubt in the first meditation. Though it is necessary that the physical world exist, he continued, there is no corresponding necessity that it be in any way similar to the world our senses depict.[3] No other statement in the century better summarized the program of the mechanical philosophy that dedicated itself to the proposition that the underlying reality of nature is utterly different from its surface appearance. On the surface, nature presents itself to us, as virtually the unanimous tradition of Western philosophy had accepted it, in organic terms. In reality it is a complex machine; the very concept of life is an illusion. So is much else. With the exception of extension, figure, and bulk (a list that varied somewhat from person to person), all of the properties attributed to bodies in the old philosophy were also stamped as illusions, the projection of our subjective sensations onto the external world. Nature is only a complex machine. Reality consists of particles of matter in motion, and particles of matter in motion produce all of the phenomena we observe. There is no necessity and indeed no reason to suppose that nature is in any way similar to the world our senses depict.

For my purposes it is important that the mechanical philosophy also banished from existence another denizen of some previous philosophies—attractions of any kind. No scorn was too great to heap upon such notions. From one end of the century to the other, the idea of attractions, the action of one body on another with which it is not in contact, was anathema to the dominant school of natural philosophy. Galileo could not sufficiently express his amazement that Kepler had been willing to entertain the puerile notion, as he called it, that the moon causes the tides by acting upon the waters of the sea.[4] In the 1690s, Huygens and Leibniz found similar ideas just as absurd for the same reasons.[5] To speak of an attraction whenever one body was seen to approach another was to philosophize on the same plane with Molière's doctor who explained the power of opium to cause sleep by a dormative virtue it contained. "Bene, bene, bene," cried his examiners; "bene respondere," and Molière's ridicule was that of the entire century.[6] An attraction was an occult virtue, and "occult virtue" was the mechanical philosophy's ultimate term of opprobrium.

Two specific phenomena serve as examples. To the characteristic natural philosophy of the sixteenth century, what has been called Renaissance Naturalism, magnetism was the ready exemplar of the vast range of mysterious forces with which it conceived the universe to be populated. Mechanical philosophy viewed magnetism as something it must explain away if its program were to claim validity. In his *Rules for the Direction of the Mind* Descartes treated it almost as a test case. When the ordinary person confronts magnetism, he said, he immediately assumes that it is utterly unlike anything he is familiar with and hence grasps at whatever notion is most difficult. One ought rather to review things that are familiar and to find the points wherein they agree with the properties of magnets.[7] In the *Principles of Philosophy* Descartes took his own advice to heart. He listed in detail the known phenomena of magnetism, and he imagined a mechanism whereby the direct contact of moving particles accounted for everything. He described how the turning of the vortex on its axis produces tiny screw-shaped particles and similar pores in terrestrial material to receive the particles. He even pointed out how the motion of particles along the two opposite poles of the vortex must produce both left-hand and right-hand screws, corresponding to the two poles of the magnet. What is the reality behind the apparent attractions of the magnet? Streams of invisible particles that move ferrous bodies by direct contact.[8] Descartes's explanation of magnetism is one of the wildest flights of mechanistic imagination from the whole seventeenth century, and others of his age found it so. I am not aware of anyone who followed him on this topic, but every mechanical philosopher, far from rejecting his program, invented his own alternative mechanism consisting of invisible streams or of threads with grappling hooks or the like. Isaac Newton, the young mechanical philosopher in Trinity College, also turned his mind to the problem. He imagined two streams of particles that, in passing through iron, acquire distinct "odors" that make them sociable, one stream to iron and loadstone, the other to the aether that fills their pores.[9] Perhaps I was wrong to label Descartes's explanation the wildest flight of mechanistic imagination; it might be hard to judge Newton's any less wild. Fascinated by the unceasing flux of matter that such mechanisms seemed to present, he thought of tapping into it with some machine to realize the perpetual motion. In his undergraduate notebook he sketched in two different devices, mounted on axles and turned perpetually by the magnetic stream.[10]

Gravity offered a similar challenge. Aristotle had ascribed the tendency of bodies to fall to their nature as heavy bodies. More recent philosophers had attributed the fall of bodies to an attraction. Descartes derived it instead from the mechanical necessities of the vortex. Every body moving in a circle strives to recede from the center. In a plenum some can recede only if others move in the opposite direction. What is gravity? A deficiency of centrifugal force on the part of certain bodies that other bodies drive toward the center.[11] Once again, alternative mechanical explanations of gravity presented themselves, usually in terms of the descent toward the earth of some invisible matter that bears down bodies in its path. Newton's student notebook contains passages about the gravitating matter. In connection with it he thought once more of perpetual motion and imagined the possibility of a gravitational shield such that a wheel hung on a horizontal axis with half of it extending beyond the shield would turn perpetually.[12]

In 1675, in connection with a paper of observations of the colors in thin transparent films, Newton sent a *Hypothesis of Light* to the Royal Society. He had noted, he said, that the heads of certain *virtuosi* (that is, students of natural philosophy) ran to hypotheses, and he offered his as a possible aid to help them understand his theory of colors. The *Hypothesis of Light* dealt with more than light, however; it tied together various speculations into a total, mechanistic system of nature. Central to it was a universal aether that pervades all of space. Newton explained the reflection, refraction, and diffraction of light by a simple mechanism in the aether. Standing rarer in solid bodies than in free space, for example, it refracts light by the differential pressure at a refracting surface that pushes the corpuscles of light and changes their direction. Wave motions in the aether explained the phenomena of thin films. The aether also explained much beyond optics. Its condensation in the earth, for example, produces a steady downward flow and causes gravity. He described a simple experiment in static electricity in which the rubbing of a piece of glass, as he explained, rarefies the aether condensed in it, making the aether stream out and move tiny bits of paper. It is true that the *Hypothesis of Light* contained other elements that do not lend themselves to easy summary as a mechanical philosophy of nature. Nevertheless, the mechanistic elements in the essay were no less real than the other aspects, and for the moment they are the part I wish to emphasize.[13] Newton composed the *Hypothesis of Light* in 1675. Twelve years later he published the *Principia*, in which a rather different set of concepts displayed themselves. Shortly before the *Principia* appeared, Fatio de Duillier arrived in England after a stay in the Netherlands. In June 1687 he wrote to Huygens telling him about the book that Newton would soon bring out on the system of the world, a book his English friends were telling him would revolutionize natural philosophy. The English, he indicated, found him too Cartesian, but he wished that Newton had consulted Huygens on the principle of attraction that his book proposed. Huygens replied that he did not care whether Newton was a Cartesian or not "as long as he doesn't serve us up conjectures such as attractions."[14] Serve up a theory of attraction Newton did, more of a theory of attraction than Huygens could have imagined had he let his mind run freely over all that he considered most misguided. Having proposed a planetary dynamics based on the concept of centripetal attraction, Newton went on to argue that every particle of matter in the universe attracts every other particle, a theory that could not have flouted mechanical sensibilities more openly.

The *Principia* carried several disclaimers, of course. He treated centripetal forces as attractions in Section XI, he said, "though perhaps in a physical strictness they may more truly be called impulses. But these Propositions are to be considered as purely mathematical; and therefore, laying aside all physical considerations, I make use of a familiar way of speaking to make myself the more easily understood by a mathematical reader." And again: "I here use the word *attraction* in general for any endeavor whatever, made by bodies to approach to each other, whether that endeavor arise from the action of the bodies themselves, as tending to each other or agitating each other by spirits emitted; or whether it arises from the action of the ether or of the air, or of any medium whatever, whether corporeal or incorporeal, in any manner impelling bodies placed therein towards each other. In the same general sense I use the word *impulse*, not defining in this treatise the species or physical qualities of forces, but investigating the quantities and mathematical proportions of them. . . ."[15] Later on, in the second English edition of the *Opticks* he inserted a series of Queries that proposed the existence of an aether that, among other things, he supposed to cause gravity, not by streaming toward the earth and sun, but by a constant differential pressure that increases steadily with distance from bodies.[16]

For all the disclaimers, however, and despite the Queries in the *Opticks*, it is impossible to imagine a mechanism adequate to the job of universal gravitation. As he prepared the text of the second edition for the press, Roger Cotes tried to raise this problem with Newton in connection with the third law. As far as the attraction of the sun on a planet was concerned, Newton might argue that he merely used the word "attraction" without making any claim about its ontological status. One could imagine an aetherial mechanism to explain it. By the third law, however, and according to the necessities of Newton's system, the planet also attracts the sun. What aetherial mechanism would explain this?[17] Although Newton swept the objection aside, it made a valid point. Other aspects of the *Principia* reinforced it. If one is not too concerned with the niceties of the lunar orbit, he can imagine some aetherial device to explain the moon's centripetal "attraction" toward the earth. Newton also devoted some attention, however, to the moon's attraction on the earth, on the one hand raising the waters of the sea in the tides, on the other hand moving the bulge of matter about the equator and causing the earth to precess. What did Query 21 have to say about these phenomena? What did it have to say about Proposition VII of Book III of the *Principia*, "that there is a power of gravity pertaining to all bodies, proportional to the several quantities of matter which they contain," such that every particle of matter in the universe attracts every other particle? The fact is, Newton's disclaimers did not make sense. The *Principia* was about attractions, and the concept of attractions flew in the face of the orthodoxy accepted by the entire scientific community of his day. If the word "revolution" will bear one more use, surely this was a scientific revolution.

Here then is the question to which I address myself. The concept of attractions, though certainly not unknown before, represented the intrusion of a new and alien idea into the accepted body of natural philosophy. I wish to insist on the conceptual issue. I cannot myself find any way to extract the idea from, say, the mathematical necessities of the situation. Newton was not the only sophisticated mathematician of the day. Christiaan Huygens, Gottfried Wilhelm Leibniz, and Johann Bernoulli were, after Newton, the leading mathematicians of Europe at the time, in the opinion of historians rather considerable mathematicians all. Even with the *Principia* laid open before them, they were unable to find that its mathematical demonstrations led to the idea of attractions, and to their deaths not one of them accepted it. We are dealing with a conceptual issue, not a mathematical one. I pose the question: can we find the source from which Newton drew it? I pose the further, more specific question: do we have in the idea of attractions an enduring influence of alchemy on Newton's scientific thought?

Alchemical study did not go back to Newton's undergraduate days. In no real sense did chemical study go back that far. Robert Boyle was one of the authors he learned to know, however, and if he read him first as a mechanical philosopher, it was not long before he was reading him as a chemist. Sometime in the 1660s, after his undergraduate days had ended, if we can judge by the hand, he composed a glossary of chemical terms, including such things as "Acid salts spirits and juices," "Aqua fortis," and "Cementation." No odor of alchemy clings to the material that he entered under these headings; it smells rather of sober chemistry.[18] His recipe for making phosphorus also smelled of sober chemistry, and of something else besides. "Take of urine," it began, "one barrel."[19] It is true that his glossary included headings such as "Alkahest" and "Projection," but it is equally true that he did not enter anything under them. By the late 1660s his reading was changing. His accounts for 1669 show the purchase of the huge six-volume *Theatrum chemicum,* the greatest collection of alchemical texts ever published, during a visit to London, and notes from it as well as from other alchemical writers such as Basil Valentine, Sendivogius, Eirenaeus Philalethes, and Michael Maier, notes written in the hand of the late 1660s, survive.[20] On the trip to London he also purchased chemicals and two furnaces, and one of his notebooks records experiments apparently from about this time.[21] It is important to note the order of his progress. Newton did not stumble into alchemy as a youth, recognize the emptiness of its promise, and turn instead to rational chemistry. Rather, he started with rational chemistry (insofar as a distinction between "rational" chemistry and alchemy is valid for the seventeenth century) and worked his way into alchemy. The man who purchased the *Theatrum chemicum* in 1669 had already invented the fluxional calculus, discovered the heterogeneity of light, and entertained rudimentary hints of the law of universal gravitation. For the following twenty-five years, while he scarcely touched optics except to present work done earlier, while he turned to mathematics less and less frequently, while he ignored what he had done in mechanics until a chance visit from Halley triggered an intense investigation for two and a half years in the middle 1680s, for the following twenty-five years, I say, alchemy was his most consistent scientific activity.

Let us be clear that we are talking about an extensive application of effort to which an extensive body of surviving papers testify. I have devoted some time to a reliable quantitative measure of these papers. They appear to contain well over a million words in Newton's hand. An activity that left behind a record that large was clearly more than an incidental diversion.

The papers fall into various categories. Many of them are reading notes. It is frequently asserted that all of them are reading notes, so that from them one cannot conclude anything with assurance about Newton's own attitude toward the subject. This is simply not true. Meanwhile, the reading notes themselves cannot be dismissed that lightly since they bear testimony to study of the alchemical tradition that was both intensive and extensive. Eighty-seven closely written pages, for example, explored the works of Michael Maier.[22] Newton was always concerned to compare one alchemist with another in order to winnow the grain from the chaff. Early in the 1680s he began to compile an *Index chemicus,* as he called it, that would help him to control the body of information he was compiling.

The *Index chemicus* advanced through three successive stages. It began with 115 headings entered on two sheets folded twice to give him eight folios in all, and under the headings he entered page references to places where he could find those topics discussed. He also squeezed new headings in until he exhausted the space available and started anew on twenty-four folios, now with 251 headings. The second version expanded as the first had done, reaching 714 headings before it could hold no more. It also began to change its character. Initially the *Index chemicus* appears to have been a typical Newtonian device to organize and command his information. He invariably began every new study with something similar. Some of the entries began to take on the form of short essays, however, and as he started to expand the *Index* anew, Newton devoted some effort to drafts of the essays on separate sheets. He started a new version, abandoned it before he completed the letter A, and treated its material as further drafts. The second version of the *Index* contained a few references to Mundanus, as Newton called Edmund Dickinson's letter to Theodore Mundanus published in 1686, references always squeezed onto the ends of lines. While it is impossible to know exactly when he read Mundanus, it is reasonable to date the beginning of the final version of the *Index* about 1686 or 1687. This final version, final in the sense that none ever replaced it, though not final in the sense that it was ever completed, stretched beyond 100 pages and contained 879 headings. I analyzed the content of the forty-six largest entries, which together fill about 42 pages and thus constitute a sizable proportion of the whole. I counted 1,975 separate page references to 141 separate treatises and 104 different authors. By extrapolation, the entire *Index* must contain about 5,000 page references, undoubtedly with several more authors not represented in the longest entries. The *Index* reveals that Newton had consumed the contents of all the major collections of alchemical works, *Theatrum chemicum*, *Artis auriferae*, *Musaeum hermeticum*, *Theatrum chemicum Brittanicum*, and *Aurifontina chymica*. He had digested as well the various works of all the major figures in the long history of alchemy—men such as Arnold of Villanova, Raymond Lull, Nicholas Flammel, Johann Grasshoff, Michael Maier, and the pseudonymous Eirenaeus Philalethes, an English alchemist still alive when Newton took up the Art and the master of it who influenced him the most.[23] Newton collected as he read. More than 10 percent of his library at the time of his death consisted of alchemical works; since he hardly collected during his final twenty-five years, the proportion at one time must have been far greater.[24] Although I am not really in a position to judge, I am willing to venture the opinion that alchemy has never had a student more widely and deeply versed in its sources.

One peculiar feature of his alchemical studies deserves mention. Among his papers is a large sheaf of alchemical essays copied in several hands, almost all essays that have never been published. In the sheaf are one paragraph and several corrections in Newton's hand of the mid or late 1660s, and elsewhere notes that he made from the collection survive.[25] The notes suggest to me that he borrowed this collection from someone to study and for whatever reason failed to return it. He also possessed a number of copies, in his own hand, of other alchemical tracts that have never been published. His writing in these copies indicates that they stem, not from any one period, but from the whole extent of his career in alchemy. His copy of one states that "Mr. F" (in the Cambridge context probably Magister F, an academic colleague) gave it to him in 1675. One Ezekiel Foxcroft, a member of the alchemical circle that had gathered initially around Samuel Hartlib, lived in King's College at this time; Newton later studied his translation of the Rosicrucian *Chymical Wedding* and referred to it as Mr. F's work. The manuscripts of Eirenaeus Philalethes circulated originally through the Hartlib circle to which Foxcroft belonged; Newton had access to some of them about ten years before their publication. In 1683 he received a letter solely concerned with the Art from one Fran. Meheux, and in the 1690s, scarcely a month before he accepted a position that put him in charge of His Majesty's coinage in gold and silver, an unnamed alchemist looked him up in Cambridge to discourse about the Work.[26] I see no way to interpret these various facts except to conclude that Newton was in touch with clandestine alchemical circles from which he received unpublished alchemical literature. It seems very likely to me that he also fed material into the network. His name would not have appeared on such papers. Until very recently those familiar with alchemical materials have not exploited Newton's manuscripts and have not been familiar with his handwriting. I shall be surprised if we do not begin soon to discover hitherto unknown Newtonian writings on the Art.

It has long been known that Newton left a large batch of alchemical papers behind. As I mentioned, it has most frequently been asserted, incorrectly, that they consist entirely of reading notes. Quite a few of them are Newton's own compositions. Rather early in his studies he drafted a somewhat disjointed paper, usually called *The Vegetation of Metals*, that contains a long essay on alchemical themes. In the late 1670s he drew upon his own experimentation in a paper called *The Key*, which Mrs. Dobbs has explicated with great success, and about the same time another piece also closely related to experience in the laboratory on *The Separation of Elements*. He wrote a commentary on the *Tabula smaragdina*.[27] He was much concerned to draw the alchemical authorities, which he knew so well, together into one consistent statement of the Work. There is a sheet with seventeen titles that sound like the titles of chapters, and elsewhere drafts of chapters, filled with material drawn from alchemical writers, with titles that correspond to those on the list.[28] A paper called *The Regimen* sums up the Work in seven "Aphorisms," the process, he stated, as he found it in "ye work of the best Authors, Hermes, Turba, Morien, Artephius, Abraham ye Jew and Flammel, Scala, Ripley, Maier, the great Rosary, Charnock, Trevisan. Philaletha. Despagnet." Four long pages supported the aphorisms with citations from these authors.[29] Another somewhat later compilation started with the title *Decoctio* but shifted in a later version to the name of the earlier paper, *The Regimen.*[30] *The Method of ye work*, originally a comparison of Didier's work with that of eleven other authors, went through a second stage that concerned itself principally with Ripley, and ended as a paper that exists in two drafts, *Praxis*, arguably the most important alchemical paper Newton ever wrote and one that explicitly claims to have achieved multiplications.[31] The most extensive of his compilations, which carried no title, was divided into a number of chapters, or "opera." We have six drafts of one of the opera, four of another, at least two of three more.[32] In many of these papers, especially the last one, most of the content consists of citations from others. Newton must have known his favorite authors almost by heart. In some of the papers he left empty parentheses after citations where he could later fill in page numbers to the passages in question. Nevertheless, these were his own compilations, testifying to his conviction that all true alchemists pursued the same Great Work. The papers were the result of extensive study and thought. There is no way to dismiss them as reading notes.

Finally, Newton pursued alchemy in the laboratory as well. Notes of his own experiments are among his earliest chemical papers. When Humphrey Newton was with him, from 1683 until 1688, he found him experimenting in the garden laboratory outside his chamber in Trinity; Humphrey's most vivid recollection of those years centered on the laboratory.[33] Recently Mrs. Dobbs has correlated Newton's early experiments with his alchemical papers and proved to my satisfaction that we can only call the experimentation alchemical. Experimental records continue through the 1680s and into the 1690s. They still await final explication, but the interjection among sober laboratory records of comments such as "I saw Sophic Sal Ammoniac" and "I made Jupiter fly on the wings of the eagle" serve to convince me at least that this experimentation too, like the earlier, was alchemical.[34] The last dated experimental notes come from February 1696, less than a month before the Lords Commissioners of the Treasury appointed Newton Warden of the Mint.[35] One can only surmise that they had not heard what was going on in the garden next to the great gate of Trinity.

Newton's alchemical papers have been a source of anguish to many scientists to whom they seem the negation of everything to which they would assign the name science and thus of everything for which Newton stands in their eyes. Newton's great biographer in the nineteenth century, Sir David Brewster, could scarcely bring himself to confess that his hero had not only copied "the most contemptible alchemical poetry" but had even annotated "the obvious production of a fool and a knave." Indeed, the only solace Brewster could extract from the situation, and he clearly found it little enough, was the fact that Leibniz had also dabbled in the Art.[36] Perhaps we are no longer so strict in our notions of proper scientific behavior. At least in recent years an increasing number of scholars have begun to look seriously at the papers. Their interpretation will probably remain a matter of contention, but a few facts have become clear. The papers are undoubtedly authentic. As I indicated, they are quite extensive. They are the products neither of Newton's youth nor of his age, but of the middle years of his life when he was at the height of his powers, the years when he completed both of the books on which his enduring reputation in science rests. To me at least it appears obvious that any activity that succeeded in absorbing that much of his time and energy must have held great meaning for him. The chronology involved, spanning the time when he arrived at the concept of attraction with which he revolutionized natural philosophy, seems to invite us to explore the possibility that the alchemical tradition could have offered a source from which he drew, not the idea, but stimulus toward it.

Compared to the mechanical philosophy, alchemy embraced a radically different view of nature. Where the one denied the reality of life and looked upon animals and plants as complicated machines, the other extended the organic outlook over every facet of nature, including the generation of metals that it believed grew in the earth. As I mentioned, what was probably Newton's earliest independent effort in alchemy was a paper on the vegetation of metals. Especially, alchemy asserted that all things and above all, metals, are generated by the union of male and female. One of Newton's favorite materials for experimentation in the 1680s was a substance he called the net, a regulus made from antimony, copper, and iron. In the imagery of alchemy Venus represented copper and Mars iron. The net referred to the story in ancient myth of a net of gold woven by Vulcan, the husband of Venus, who had good cause to suspect both Mars' intentions and his wife's reception of them, and with the net captured the pair *in flagrante delicto*. One might search some time to find a more suggestive image than the net; it represented the alchemical hermaphrodite, the union of the male and female principles. To the image of sexual generation alchemy frequently joined a second image of purification. Before they could join in effective union, both male and female, animating soul and animated body, had to be purged of the encumbering feces that weighed them down. Alchemists purified their materials eternally; much of the practical work in the laboratory concerned itself with this task. On occasion, the juncture of the two themes, generation by male and female and purification, could produce some striking imagery. "I say our true Sperm flows from a Trinity of Substances in one Essence," Philalethes wrote, "of which two are extracted out of the Earth of their Nativity by the third, and then become a pure milky Virgin-like Nature drawn from the Menstruum of our Sordid Whore." Having seen and duly marked this passage, I found that Newton had discovered it before me and liked it well enough to repeat it at least seven times in his papers.[37]

Associated with its organic conception of nature, alchemy also thought in terms of active principles, embodied most perfectly in the philosophic sulfur (inevitably male in that age), which animated the passive (female) matrix, the philosophic mercury. To the mechanical philosophy active principles were anathema. Matter in its view was wholly inert, incapable of initiating any activity or change, subject to being moved wherever an external impulse directed it. Chemistry almost cried out aloud against the effort to reduce it to such mechanical terms. Repeatedly its phenomena seemed to reveal foci of activity in matter. Two cold materials gently mixed suddenly grew hot. Two substances joined together to form a compound but shunned a third. It is striking how readily active verbs inserted themselves into Newton's descriptions of experiments. An acid solution "wrought upon" spelter until it dissolved it, and another solution "fell a working wth a sudden violent fermentation." Spirits "drew" or "extracted" the salts from metals. When substances combined, one "laid hold" on the other; when two sublimed together, one "carried up" the other, and if they did not sublime, it was because one "held" the other "down."[38] Phenomena such as these gave great trouble to mechanistic philosophers forced to explain all of nature with only the categories of size, shape, and motion. In his alchemical reading Newton constantly met the concept of active principles in nature, and in his laboratory he constantly witnessed phenomena that did seem to require them.

From an early time the speculations on the nature of things by Newton, who had begun his scientific career by conversion to the mechanical philosophy, took on a special appearance as he wove the concept of an active–passive dichotomy into them. In the late 1660s his essay on *The Vegetation of Metals* distinguished nature's two modes of action, the vegetable and the mechanical, in terms of the active–passive dichotomy. The principles of nature's vegetable actions are seeds, "her only agents, her fire, her soule, her life." A seed is never more than a tiny part of the whole surrounded by "dead earth & insipid water." The grosser substances act as vehicles only, and they take on different appearances as their particles are moved about. These changes, Newton argued, are purely mechanical. One can get a different hue from the mixture of two colored powders, for example, or one can coagulate milk into butter by agitating it. Although vulgar chemistry can produce impressive displays, all of her operations involve nothing but the conjunction and separation of particles, that is, mechanical alterations, and nature employs the same means to produce the same effects.

> But so far as by vegetation such changes are wrought as cannot be done wthout it wee must have recourse to som further cause And this difference is vast & fundamental because nothing could ever yet be made wthout vegetation wch nature useth to produce by it. . . . There is therefore besides ye sensible changes wrought in ye textures of ye grosser matter a more subtile secret & noble way of working in all vegetation which makes the products distinct from all others & ye immediate seat of thes operations is not ye whole bulk of matter, but rather an exceeding subtile & inimaginably small portion of matter diffused through the mass wch if it were seperated there would remain but a dead & inactive earth.[39]

The active–passive dichotomy expressed itself in the "Hypothesis of Light" in 1675, both by presenting the aether as the active spirit of nature and by dividing the aether itself into the "maine flegmatic body" and "other various aethereall Spirits. . . ."[40] Moreover, when Newton began to work toward the *Principia* in the autumn of 1684, his early essay *De motu* did not embrace the principle of inertia but rather presented the motions of bodies as the results of the interactions of forces external to bodies with forces internal to them, that is, with active principles animating lifeless matter that plays a wholly passive role.[41]

Two and a half years later, when he composed the preface for his now completed *Principia*, Newton briefly described the content of the three books and his use of the force of gravity in Book III to explain celestial motions. "I wish we could derive the rest of the phenomena of Nature by the same kind of reasoning from mechanical principles," he continued, "for I am induced by many reasons to suspect that they may all depend upon certain forces by which the particles of bodies, by some cause hitherto unknown, are either mutually impelled towards one another, and cohere in regular figures, or are repelled and recede from one another."[42] At one point he had planned to say much more about these other forces. He had drawn up an extended "Conclusio" to his work that described at some length the evidence for the inter-particulate forces. "Hitherto I have explained the System of this visible world," it began, "as far as concerns the greater motions which can easily be detected. There are however innumerable other local motions which on account of the minuteness of the moving particles cannot be detected, such as the motions of the particles in hot bodies, in fermenting bodies, in putrescent bodies, in growing bodies, in the organs of sensation and so forth. If any one shall have the good fortune to discover all these, I might almost say that he will have laid bare the whole nature of bodies so far as the mechanical causes of things are concerned."[43]

To support this assertion Newton drew primarily upon the evidence of chemistry. He was greatly impressed by reactions that produce heat.

> If spirit of vitriol (which consists of common water and an acid spirit) be mixed with Sal Alkali or with some suitable metallic powder, at once commotion and violent ebullition occur. And a great heat is often generated in such operations. That motion and the heat then produced argue that there is a vehement rushing together of the acid particles and the other particles, whether metallic or of Sal Alkali; and the rushing together of the particles with violence could not happen unless the particles begin to approach one another before they touch one another. . . . So also spirit of nitre (which is composed of water and an acid Spirit) violently unites with salt of tartar; then, although the spirit by itself can be distilled in a gently heated bath, nevertheless it cannot be separated from the salt of tartar except by a vehement fire.

In addition to reactions in which heat appears in formerly cold ingredients, reactions displaying selective affinities also suggested forces between particles. When bodies dissolved in acids are precipitated by salt of tartar, he argued, "the precipitation is probably caused by the stronger attraction by which the salt of tartar draws those acid spirits from the dissolved bodies to itself. For if the spirit does not suffice to retain them both, it will cohere with that which attracts more strongly."[44]

Newton did not discover these reactions; they were well known to the chemists of his day. All or virtually all of them appeared, for example, in the writings of Robert Boyle, with which Newton was familiar. What Boyle fitted into the corpus of his mechanical chemistry became for Newton evidence of active principles by which particles of matter attract or repel each other. He could not have found this conclusion in Boyle. Nor could he have found it, in the form in which he cast it, in alchemical writers. In them he could have found a concept of active principles associated with analogous phenomena, however. Suffice it to say that without exception the chemical phenomena cited in his "Conclusio" of 1687 appeared among his alchemical papers of the previous decade.

In the *Principia* itself, the active–passive dichotomy, recast from its original form reappeared in the duality of impressed forces external to bodies and internal *vires inertiae* whereby they resist the effort of external forces to alter their state of rest or motion. In Query 31 of the *Opticks* Newton explicitly drew out the active–passive imagery behind his conception.

> The vis interiae is a passive Principle by which Bodies persist in their Motion or Rest, receive Motion in proportion to the Force impressing it, and resist as much as they are resisted. By this Principle alone there never could have been any Motion in the World. Some other Principle was necessary for putting Bodies into Motion; and now they are in Motion some other Principle is necessary for conserving the Motion.

Newton continued with the argument by showing, first, from the composition of motions, that the total quantity of motion is not universally conserved and, second, that motion is constantly lost through friction and imperfect elasticity.

> Seeing therefore the variety of Motion which we find in the World is always decreasing, there is a necessity of conserving and recruiting it by active Principles, such as are the cause of Gravity, by which Planets and Comets keep their Motions in their Orbs, and Bodies acquire great Motion in falling; and the cause of Fermentation, by which the Heart and Blood of Animals are kept in perpetual Motion and Heat; the inward Parts of the Earth are constantly warm'd, and in some places grow very hot, Bodies burn and shine, Mountains take fire, the Caverns of the Earth are blown up, and the Sun continues violently hot and lucid, and warms all things by his Light. For we meet with very little Motion in the World, besides what is owing to these active Principles. And if it were not for these Principles, the Bodies of the Earth, Planets, Comets, Sun, and all things in them, would grow cold and freeze, and become inactive Masses; and all Putrefaction, Generation, Vegetation and Life would cease, and the Planets and Comets would not remain in their Orbs.[45]

While I find the references to putrefaction, generation, vegetation, and life suggestive, there are other passages in which I perceive a more direct filiation between the Newtonian conception of force and alchemical ideas. Newton held that the refraction and reflection of light are caused by forces exerted by refracting and reflecting media upon corpuscles of light. In the *Principia* he demonstrated how an attraction normal to a refracting interface entails the sine law of refraction, and in the *Opticks* he drew up a table relating the refractive indices of various media to their specific gravities.[46] As he noted, there appeared to be two distinct classes of refracting bodies, those that abound in "sulphureous oily Particles" and those that do not. "Whence it seems rational," he concluded, "to attribute the refractive Power of all Bodies chiefly, if not wholly, to the sulphureous Parts with which they abound. For it's probable that all Bodies abound more or less with Sulphurs. And as Light congregated by a Burning-glass acts most upon sulphureous Bodies, to turn them into Fire and Flame; so, since all Action is mutual, Sulphurs ought to act most upon Light."[47] In alchemy, sulphur was the ultimate active agent. It appears to have retained that role in Newton's universe.

De natura acidorum, a paper written in the 1690s related sulphur to another of the active substances of the alchemical world, acids, the dragons and serpents that devoured sundry kings and queens in numerous alchemical texts. In this case Newton suggested that the activity of sulphur, presumably common rather than philosophic sulphur, derives from the acid it contains. "For whatever doth strongly attract, and is strongly attracted, may be call'd an Acid." One sentence in *De natura acidorum* pointed at the role of the alchemical active principle in the Newtonian conception of force. "The Particles of Acids . . . ," he asserted, "are endued with a great Attractive Force; in which Force their Activity consists; and thereby also they affect and stimulate the Organ of Taste, and dissolve such Bodies as they can come at."[48] One offspring of the forces between particles was a radical new conception of body that reduced it largely to empty space sparsely seasoned with diaphanous threads of matter. Significantly, he drew upon his alchemical material to describe it by the image of a net.[49]

I trust that my argument will not be misinterpreted. I am not contending that Newtonian science is a disguised form of alchemy. Nor am I asserting the more confined argument that Newtonian attraction is merely an alchemical active principle. Rather, I am seeking to find a source that might have stimulated Newton's mind toward what became at once his break with the prevailing orthodoxy of seventeenth-century science and the concept that enabled him to carry that science to its highest level of achievement. Newton started his career within the school of mechanical philosophy that rejected any idea of attraction as an occult throwback to a misguided past. Such an idea formed the very core of his *Principia*. I argue that in alchemy Newton found a school of natural philosophy that embodied concepts alien to the mechanical philosophy, and specifically the concept of active principles that animate passive matter. Alchemical active principles were not Newtonian forces, however. Whatever we do, we must not treat one of history's rare giants as the passive recipient of others' ideas. Above all else Newton himself was an active principle who reshaped what he received into products that were his own. He exercised his alchemy on alchemy itself. The idea of attraction did show up in some alchemical writers, including Sendivogius, whom Newton read carefully.[50] Nevertheless, the general idea of attraction is not Newtonian force. It had not yet acquired that exact mathematical definition that made the concept not an occult notion, but the heart of a powerful quantitative science. This Newton gave it, transforming the alchemical notion at least as much as the notion itself transformed mechanical philosophy.

There has been considerable research in the last few years on the influence of the Cambridge Platonists, especially Henry More, on Newton, especially in connection with the concept of active principles.[51] I do not see that what I propose now conflicts with that research. There is no reason why a number of sources known to have influenced Newton could not work in rough harmony toward the same end. Let me only insist that the evidence of Newton's involvement in alchemy is not conjectural. It exists in the form of more than a million words in his hand. In contrast, his notes on More and Cambridge Platonists do not stretch even close to a tenth of this amount. Whatever the influence of the Cambridge Platonists, Newton's career in alchemy was real and extensive and, the papers indicate, reached its peak at the very time of the *Principia*.

From the point of view of alchemy, the *Principia* was an interruption. Indeed, Newton paused in the spring of 1686, when he had not yet put either Book II or Book III into its final form, to perform a series of alchemical experiments.[52] And once the book was out, he returned both to his manuscripts and to his laboratory with undiminished intensity. I find in the continuation of alchemical research in the early 1690s powerful confirming evidence that Newton saw the Art in harmony rather than in conflict with his masterpiece. As far as I can judge from the hand, about half of his alchemical papers come from the five-year period following the *Principia's* publication.

Among other things, Newton introduced Fatio de Duillier, the young Swiss mathematician who entered intensely into his life at this time, to the Art, and both Fatio and alchemy figured in the mounting emotional tension that wracked Newton's life in 1693. In the summer of that year, sometime after his reception of Fatio's letter of May 13th that he cited, Newton composed the essay *Praxis*, which I have called possibly his most important alchemical paper. At the climax of *Praxis*, Newton claimed to have achieved multiplication.

Thus you may multiply each stone 4 times & no more for they will then become oyles shining in ye dark & fit for magicall uses. You may ferment them with ⊙ and ☽ by keeping the stone & metall in fusion together for a day, & then project upon metalls. This is the multiplication in quality. You may multiply it in quantity by the mercuries of wch you made it at first, amalgaming ye stone with y^e ☿ of 3 or more eagles & adding their weight of ye water, & if you designe it for metalls you may melt every time three parts of ☉ wth one of ye stone. Every multiplication will encrease its vertue ten times & if you use y^e ☿ of ye 2^d or 3^d rotations wthout ye spirit perhaps a thousand times. Thus you may multiply to infinity.[53]

I do not think that we should take this passage seriously, certainly not as evidence that Newton attained the alchemist's goal, but no more as evidence that he believed he had. Newton was overwrought at the time. Within the following three months he wrote the famous letters to Locke and Pepys, which both of them took as evidence of some derangement, as everyone who reads them must. Rather, I offer the passage as an indication that twenty-five years of deep involvement in the Art reached their culmination in what was also, for a number of reasons, the tragic climax of Newton's life.

From dated experiments we know that Newton continued to work in his laboratory after 1693. Although most of the papers are not dated and cannot be placed with precision, there is no reason to think that some of them do not also come from the years immediately following the breakdown. There are at least four scraps related to business at the Mint during his early tenure there.[54] Sometime between 1701 and 1705 he purchased a number of alchemical works along with other books.[55] Nevertheless, we cannot avoid the fact that Newton did turn away from alchemy. If, as I just noted, there are four scraps with Mint business on them, the fact that there are only four is significant. Only three of the alchemical books in his library have an imprint after 1700, and two of those, bound together, were the gift of their author, William Yworth, to whose support, it appears, Newton did contribute.[56] Of the major intellectual pursuits of his Cambridge years, alchemy alone failed to follow him to London to participate in the final chapter of his life. Only recently has the full import of this fact impressed itself upon me, and I wish to suggest that his turning away from alchemy is a reality of no less importance than his involvement in it in the first place. Unfortunately he left no single word, as far as I have been able to find, to explain his decision. One is forced back entirely upon speculation. Since I have been trying to insist in this paper that I am speaking from solid evidence and not from tenuous hints, I want to insist doubly that what is purely speculation be labeled as such. Two possible reasons, not mutually exclusive, for Newton's final rejection of alchemy occur to me. First, he may indeed have become disillusioned with the Art. When he returned to some semblance of normality in the autumn of 1693, the man who prided himself on separating demonstrations from mere hypotheses may have been shocked to read what he had claimed in the *Praxis*. The manic exaltation of the summer had apparently included an alchemical dimension as he seemed finally ready to embrace the Venus of his dreams, Truth in all its seductive appeal. In the end Newton's whole active career in science dissolved away in the following depression; the alchemical dream may well have dissolved with it. But I would suggest as well a second, related possibility. As the full measure of his achievement in the *Principia* was borne in upon him in his new role as doyen of British science, Newton may have realized that in fact he had embraced Truth, not quite the enchanting Venus of his alchemical dream, but a very satisfactory mistress nevertheless. He had extracted the essence of alchemy; the Art itself counseled him to reject the dross that remained. With the quantified concept of force he had set science on a new track. Alchemist in one sense to the end, he projected that concept into the ready soil of natural philosophy. It has continued to multiply since then beyond even his wildest dreams.

SIX

Biology and Behavior— A Modern Example of Paradigm Conflict

George R. Terrell

For as long as there have been records of human speculation on the nature of the world one of the primary topics of that speculation has been human nature itself. What aspects of behavior distinguish people from animals and from the various superior beings of mythology? One aspect of that question that has often been argued in the past and continues to be a source of controversy is the so-called nature–nurture debate. This asks, to what extent do behavioral differences among humans arise from influences outside the organism (environment), and to what extent do they follow from biological individuality (heredity)? In this paper certain principles from the history and philosophy of science will be invoked to explain the limited progress toward resolution of this controversy, even with the increasingly powerful tools of modern natural science at hand. In the process it will be argued that the division is not between hereditarians and environmentalists at all; and a more appropriate characterization of the split will be proposed.

Francis Galton was born in 1822 in Birmingham, England, the son of a well-to-do banker; this circumstance made it possible for him to live the life of one of those Victorian dilettantes whose range of accomplishment so impresses us today. In his twenties, while traveling in South-west Africa, he single-handedly stopped a tribal war with little more than English imperialist chutzpah. In his long life he was a geographer, one of the founders of systematic meteorology, a pioneer in making fingerprint identification practical, and a grandfather of modern statistics.[1]

A major watershed in his life was the publication by an admired older cousin of a book called *The Origin of Species.*[2] Charles Darwin had presented there a theory of the present great variety of life on Earth based on two fundamental precepts: that all living creatures have descended from a common ancestor; and that the gradual, evolutionary change leading to the observed diversity of the biosphere was driven by natural selection of those individuals successful at reproduction. This work has been at the heart of all subsequent work in the history and classification of living things.

Galton seems to have been inspired by the possibility of applying Darwin's ideas to human beings. After all, anatomy and physiology strongly suggested that man was part of the animal kingdom; thus, the theory of evolution would be expected to apply to human beings as well. In fact, Darwin himself published in 1869 a book called *The Descent of Man,*[3] which explicitly included Homo sapiens in the evolutionary scheme. At any rate, a considerable portion of the rest of Francis Galton's life was devoted to measuring people in various ways—physical, social, and psychological—in order to understand the natural variety on which natural selection must have acted if indeed the theory were relevant to man.

In 1869 Galton published a readable and thought-provoking book called *Hereditary Genius: An Inquiry into Its Laws and Consequences.*[4] In it he set out to identify a large number of men of great accomplishment in a variety of fields of endeavor, including judges, statesmen, commanders, writers, scientists, poets, musicians, painters, clerics, scholars, oarsmen, and wrestlers. His purpose was to ascertain as far as possible their family connections. He came to two major conclusions: that close relatives of eminent men are far more likely to achieve eminence than are people in general, and that this advantage rapidly declines as one's relationship to the eminent man becomes more tenuous. Such results are at least compatible with the possibility that eminence in certain fields is influenced by personal characteristics that vary from person to person and are inherited. Furthermore, the work is a treasure trove of methodological advances and interesting side observations. For example, Galton has worked out a plausible explanation for his impression that the children of clerics often turn out badly.[5] The book is clearly a sizable contribution to several areas of scholarship.

Yet there is a mystery here. More than a century after its first publication, the book remains highly controversial. For example, Stephen

Gould, contemporary zoologist and popularizer of science, in his best-selling critique of intelligence testing, *The Mismeasure of Man*, warns us that the book should not be "mistaken for the harmless musings of some dotty Victorian eccentric"; and he repeatedly places Galton in his rogues' gallery of victims of irrational hereditarian bias.[6] It is easy to find many other instances of modern scientists who completely reject Galton, and equally easy to find others who, in relation to the same work, consider him the brilliant founder of a vast domain of inquiry. Many of us have believed naively that science itself evolves and that a reasonable time (in this case, more than a century) should be enough for thoughtful people to evaluate the work of the past, either rejecting lines of inquiry or accepting their essential truth. This clearly has not happened to Galton's work, and we are entitled to ask why.

A digression is in order. In 1962 the historian of science Thomas Kuhn published his famous treatise, *The Structure of Scientific Revolutions*.[7] In a relatively short time this work has come to loom very large in the minds of historians and philosophers of science. Almost every recent writer in either discipline feels obligated to interpret his own work in the light of Kuhn's or to deny its relevance at some length. Kuhn argues that the history of science has not been steady and evolutionary in the sense of Darwin; rather there have occurred at irregular intervals relatively abrupt, revolutionary transformations. To cite two examples among many, the intellectual dislocation generated throughout the Western world by Copernicus' theory that the sun and not the earth is the center of the solar system is well known; the similar trauma that accompanied the invention of quantum mechanics in the 1920s is less so. Kuhn believes such events to be universal in science.

To briefly summarize his theory, Kuhn asserts that the world of possible phenomena encountered by experiment and observation is so large as to be quite impossible for the human mind to encompass. Howard Resnikoff tells us that the eyes take in far more information than the visual cortex can process;[8] apparently a substantial amount of filtering and preprocessing must go on between the eyes and the conscious mind to enable us to interpret what we see. Similarly, Kuhn proposes that scientists use "philosophical preprocessors" to select and edit those observations, experiments, and styles of explanation to which their scientific work will be addressed. Kuhn calls these sets of informal limits on domains of inquiry "paradigms."

In my experiences with scientific workers in a variety of fields, good scientists seem rarely to be very open-minded people so far as their work is concerned. An excess of tolerance for new data would seem to be a serious obstruction to that process of systematization and interpretation that Kuhn calls "normal science." He goes on to describe the day-to-day labors of scientists as puzzle-solving. The challenge that keeps even exceedingly intelligent people at their scientific tasks is the constant effort to find ways to fit ever wider varieties of phenomena into the accepted paradigm. What we would normally call imagination has a limited role to play here. Apparently, the primary drive for people in these areas is to handle a limited domain thoroughly.

Finally, even a very useful paradigm, such as the belief in classical physics that every isolated system can be perfectly understood and perfectly predicted by solving the appropriate partial differential equation, eventually breaks down. Data will have been admitted that cannot readily be systematized under the existing set of rules of interpretation. By 1925 enough had been learned about the behavior of subatomic particles that it had become clear to leading physicists that something was wrong with the classical paradigm. At that point, a theory called quantum mechanics was developed, and a paradigm emerged called the Copenhagen interpretation, which posited that an observer necessarily had an effect on what he was observing. The classical isolated system was gone forever.

Given Kuhn's picture of the ideal, rather hard-headed scientist, it is not surprising to learn that such transitions can be rather traumatic. The rules of one's lifelong favorite game have been radically changed. Histories of twentieth-century science often note that Albert Einstein failed to make the transition to the quantum paradigm. He was unable or unwilling to make further contributions to particle physics after it was generally accepted among physicists that such work must be expressed in the language and mathematics of quantum mechanics. It is less well known that he was at the very least one of its godfathers—he discovered the quantization of the interaction of light and matter, and his later research into the gaslike behavior of photons led to Schroedinger's wave mechanics.[9] Further, Einstein came fairly rapidly to accept the accuracy of the predictions of the new theory. What he could never accept was the new philosophy, the Copenhagen interpretation. Apparently, good science cannot be done without a deep personal commitment to the philosophical basis of the science in question. It is interesting to note that Schroedinger himself followed Einstein into self-imposed intellectual exile, waiting for the return of the true king, classical causality.[10] Kuhn tells us that these (to mix our metaphors) Moseses who can never enter the promised land of the new paradigm are very common in the history of science.

Particularly relevant to our concerns is the problem that arises when followers of distinct paradigms try to discuss similar phenomena with each other. They rapidly find themselves talking past each other, and tempers have been known to flare. They have different vocabularies, different explanatory devices, and perhaps most important, a completely different sense of which aspects of a set of observations are most important. For example, in the early nineteenth century Dalton first proposed the theory that the rules for chemical reactions could be best explained by assuming that pure elements were made of a great many identical individual particles called atoms that were characteristic of the element. When elements combined by chemical reaction into compounds, the atoms joined in fixed numbers per type into still tiny aggregations called molecules; this immediately explained the constant proportions of various elements that were known to enter into compounds. It must be realized that at that time atoms and molecules were very much abstract concepts and part of the arcane vocabulary of the followers of Dalton. One cannot see an atom, and only recently has it been possible to photograph one.[11] Thus, a chemist who believed in them would have great difficulty talking to one who did not about the simplest chemical experiments with which both were thoroughly familiar. The first chemist would find that his very use of words like "atom," "molecule," or "valence" would obstruct communication with someone for whom these words had no referents.

Kuhn notes the tendency of scientists to portray predecessors who did not hold the current paradigm as fools, charlatans, bumbling incompetents, and so forth.[12] Often these include some of the brightest and most knowledgeable people of the day! One would hope that a degree of self-consciousness of this process would alleviate some of the more embarrassing excesses of public controversies. Let us investigate in this light the paradigmatic rift between Galton and his followers and Gould and his colleagues.

The story begins with the most important revolution in the history of biology, initiated by the appearance of Darwin's *Origin of Species*. In fact, the book was more important for the general perception of biology than it was for the science, at least at first. Among biologists, theories of evolution of life had been "in the air" for more than half a century; Darwin's painstaking defense of a principle of common descent was accepted almost immediately by those without a religious axe to grind. Much more problematic was another crucial tenet of Darwin's theory (as Ernst Mayr argues in his highly recommended book, *The Growth of Biological Thought*[13]). Evolutionary change is driven by spontaneous variation followed by differential success at reproduction. Darwin and Alfred Russel Wallace, the co-inventor of the idea, seem to have been in a small minority of biologists who accepted "survival of the fittest" in the decades that followed. For one thing, the theory was very hard to test; for another, it held very difficult technical and philosophical problems. The technical problem was that the mechanism of reproduction was not yet known; only a scientific breakthrough could resolve it.

This breakthrough was a short time coming. In 1866 Gregor Mendel published a paper announcing that he had discovered (as a result of painstaking experiments with the cultivation of peas) the principles of segregation and recombination: each creature has a great number of loci, each with specialized functions in the genetic growth program. At each site one allele (genetic instruction) selected at random comes from each parent's pair.[14] This immediately explains why children differ from parents and siblings differ from each other—they have unpredictable combinations of parental traits. Further, some alleles are recessive: they fail to influence growth very much when paired with a different instruction at the same site. This last fact is very important; Darwin had failed to explain how enough random variation could occur at convenient times to generate the enormous diversity we see about us. With recessive traits, variation could be collected by stable populations over long time intervals and then manifested as a result of inbreeding when the population dropped in time of crises. Many of the difficulties with Darwinism had been elegantly resolved.

Inexplicably, nobody influential seems to have read Mendel's paper. Sinking without a trace, it was rediscovered a third of a century later, in 1900. One might speculate that Darwin may have known as much about pigeon brooding as Mendel did about peas, but Darwin admitted himself to be quite unmathematical.[15] Perhaps a person with a mathematical aesthetic sense might have perceived the 3:1 ratio of dominant to recessive traits from heteroallelic parents (those possessing both propensities); Darwin failed to do so.

After 1900 there was an interlude while several people, including the great statistician and geneticist Ronald Fisher, worked out a satisfactory mathematics of evolutionary genetics. By 1940, the end of what Mayr calls "The Great Synthesis," virtually all biologists had accepted natural selection as the driving force behind evolution.[16] As several critics of Kuhn have pointed out, scientific revolutions may be rapid for each scientist who undergoes conversion, but they can be very slow from the point of view of the outside world.

Philosophical problems may be even greater hindrances to progress than technical problems. Darwinism says that evolution is driven by constant, enormous variation of individuals within species. No two rabbits are alike; Plato might have said that they are poor copies of an ideal Rabbit. Evolutionary theory is no longer comfortable with this—it is aesthetically and morally neutral with respect to variety. There are many different rabbits; when the environment changes unpredictably, an unanticipated subset of the original variety is likely to thrive. In all this there is no Rabbit, only lots of individual rabbits so called for the fact that they interbreed and for convenience of classification. As Mayr explains it, biology is Aristotelian in contrast to the more Platonist chemistry and physics. After all, electrons really are all alike. Mayr calls the acceptance of the importance of biological variation "population thinking," as opposed to idealistic "essentialist thinking."[17]

As the title of this paper implies, I wish here to discuss the impact of biology on the human behavioral sciences. As such, that would be too broad a topic. Besides, it would not be one story. All modern social scientists have accepted the ideas of natural selection and common descent in varying degrees. However, as we shall see, there has been only slow, piecemeal, and incomplete penetration of what Mayr calls population thinking into the behavioral sciences.

The problem is not, of course, that anyone is unaware of the great biological variation among human beings. The problem is in the application of population thinking to behavior. Variations in language, technology, and habits of cooperation have obscured to some extent any underlying biological variation in the human species in propensities to different behaviors. At least since the Enlightenment, models of human behavior of the tabula rasa sort, in which the infant human mind is thought of as a "blank slate" on which experience writes instructions, have had currency. In Darwin's time, Karl Marx proposed the principle of economic determinism, according to which adaptation to one's role in the social system was the primary determinant of behavior. Variations on this idea continue to be very influential. The later work of Sigmund Freud, especially that part emphasizing the role of traumatic childhood experiences in personality formation, also seems to be influential among social sciences. Note that innate, biological variation is only in the background in each of these models of behavior.

Yet, for the heirs of Darwin biological variation and poulation thinking are inescapable. Behavioral adaptation is at least as important to survival and reproduction as metabolic and mechanical adaptations are. It was inevitable that from the beginning Darwinian students of human behavior would search for heritable variation. Their paradigm would make it quite inexplicable if no variation was found. Thus, Galton and his heirs, prominent among them Pearson, Spearman, Fisher, Burt, Eysenck, R. B. Cattell, and Jensen, have all sought to measure genetic components in human behavior.

At this point I will make a claim that may be found surprising in light of the tone of popular discourse on the subject. The controversy over the existence of heritable behaviors is traditionally called the nature–

nurture debate, after the title of a book by Galton called *English Men of Science: Their Nature and Nurture.*[18] Both Darwin and Galton discussed the influence of the combination of heredity and environment on behavior. I have nowhere found evidence of any influential behavioral scientist of the twentieth century who has been anything resembling an exclusive genetic determinist. The Galton school should properly be called (after Burt[19]) "interactionists," because they seek to clarify the separate roles of heredity and environment in the determination of behavior. Clearly, this school finds population thinking central to their work, since biological variety and individuality are omnipresent in their theories.

In contrast, there is evidence aplenty of modern environmental determinism. Most Marxists seem to be environmental determinists, though Marx himself may not have been so rigid ("From each according to his *ability. . . .*" [my italics]). Shortly before 1920 the American psychologist John B. Watson founded the influential behaviorist school, which attempts to explain all behavior as learned (conditioned) responses to external stimuli.[20] This is, of course, an essentialist theory, positing an idealized stimulus—response mechanism in the human brain that is supposed to be universal in Homo sapiens. Behaviorism is subject to controlled laboratory experimentation, and a large body of results has accumulated in the intervening years. It is easy to imagine why an essentialist theory in opposition to the biometric school of Galton might have arisen when it did. This was the age of Einstein and the great triumphs of essentialist, laboratory-oriented physics. There we find theories positing small numbers of universal particles and forces whose ideal characteristics explain great ranges of phenomena. Many of these triumphs preceded the great synthesis in biology, so behaviorism faced a competitor in the biological study of behavior still in considerable disarray. Finally, and it is still unique in that respect, behaviorism provides a verifiable technology for behavior change. If one wants a scientific way to cure bad habits such as smoking, behavior-modification clinics are the only choices that are in any sense scientific. Thus, essentialist schools of behavior have considerable numbers of supporters and a degree of scientific success.

So we find ourselves today with a major paradigmatic gulf between behaviorist and Marxist essentialists on the one hand and poulationist biometricians (as the Galton school calls itself) and sociobiologists (as a recently organized school of genetically oriented behavioral scientists that arose out of zoology and anthropology calls itself[21]) on the other. Remember that paradigm conflicts are not arguments over the validity of data or over analysis errors. They run much deeper and are fundamentally irreconcilable. When disagreement extends to the nature of allowable explanations, very little communication is possible. You may even see complete reversals of attributed cause and effect, since each side has limited the variety of possible causes. R. D. Laing has attributed certain kinds of psychotic behavior to the fact that the victim's family situation is extremely unpleasant.[22] Other investigators point out that a psychotic is so difficult to live with that the family develops unattractive defenses. Which is cause and which is effect? Name a social problem, and there are likely to be mutually exclusive explanations of it from the two schools. Are people poor because they lack coping skills or do they lack coping skills because they are poor? Does a student do poorly in school because he is stupid, or does he seem stupid because of being failed in school? It must be remembered that attributions of cause and effect are properties of theories about phenomena and do not inhere in the phenomena themselves. Thus, we find that essentialists neglect individual differences in behavioral propensities, while populationists believe they are crucial.

We now see a reason for the Gould–Galton division. Gould seems to be a follower of the Marxist paradigm, and therefore disallows Galton's attribution of different levels of achievement to individual traits. He has nowhere experimentally disproved Galton's theories—he has simply ruled them out of order on paradigmatic grounds. It is interesting to speculate how Gould, a paleontologist, has managed to reconcile his profession with his nonevolutionary ideas about behavior. Others in a similar quandary have chosen to believe that, though biology is relevant to other aspects of human life, the brain has made a qualitative evolutionary jump beyond all other organs. It is now above the evolutionary melee and is the only organ best characterized as a Platonic ideal. Thus, the tabula rasa is reconciled with Darwin. A populationist would argue that, though the human brain is quantitatively more sophisticated than other animal brains, it is qualitatively similar, and it evolved through the same mechanisms of variation and selection. For him the same biological principles must therefore apply to behavior as to other aspects of the study of life. Naturally, he finds any number of experiments that can be given a populationist interpretation, just as his opponent finds many that can be given an essentialist interpretation.

Where is the conflict of paradigms likely to go from here? Marcus believes it has largely been resolved in cultural anthropology and that current thinking is in harmony with biology.[23] A generation ago, Cronbach diagnosed the same division among psychologists; he called it the split between the correlational psychologists (our populationists) and the experimental psychologists (our essentialists).[24] He had hope that the experimentalists were beginning to allow for individual variation and the correlationists were learning to do controlled lab work. In educational psychology and psychopathology, the population point of view seems to have carried the day. Behaviorists still possess the only technology of behavior modification, of course, so this school will continue to be important. Kenneth Bock in his recent work, *Human Nature and History*, argues cogently that biological thinking has not yet progressed far enough to make it relevant to the understanding of the history of human institutions.[25] After all, the question of how we got to the world we see around us has no ideologically pure answer; we must still study the sequence of events. All in all, it is not so bad that there are several mutually exclusive paradigms for a subject so vast as human nature. We nonscientists are free to choose those ideas that contribute the most to understanding our own experiences and problems. Perhaps that is ultimately what sciences are: competing manufacturers in the marketplace of ideas.

One rather nonscientific issue remains. My impression is that certain people whose ideas were losing the day in the race to provide cogent explanations have retreated to the arena of politics and the popular press, with considerable success. It would be tragic if the result were the suppression of good science in the name of bad. For example, Gould and others have succeeded in generating a popular impression that aptitude testing is at best a crackpot pseudoscience and at worst a vicious device for political oppression. This impression is quite false. Aptitude tests are one of the major technologies developed by the biometric school of Galton in the last century. The methods used are unexceptionable applications of the populationist paradigm (see, for example, Jensen[26]). It is interesting to note that throughout their history the primary goal of aptitude tests had been to tailor each child's education to his or her individual requirements; populationists would inevitably consider this important. Indeed, aptitude testing has come to play an important social role; it is hard to image how a college with a very intense curriculum could select those prospective students most likely to benefit from it without the aid of exams like the Scholastic Aptitude Test. The plethora of secondary schools with extremely varied standards makes a universal measure of preparation very necessary. Otherwise, such a college finds itself in the unpleasant and undemocratic position of having to reject promising students because they come from poor schools, or of accepting a large proportion of students doomed to failure. Thus, aptitude testing is a scientifically based technology with a humane and democratic social purpose.

Furthermore, the conclusion that a number of measurable human aptitudes are largely inherited is not only a possible one, it is largely inescapable for adherents to the Darwinist point of view in light of a vast quantity of evidence from many times and places.[27] The moderately toned criticism of this result has been generally from the essentialist point of view—honestly conceived but of necessity talking past its opponents. A more cogent essentialist response would be to develop a competitive set of predictors of academic achievement; so far as I know, this has not been accomplished. Watson's famous claim that he could take any five-year-old child and prepare it to succeed at any profession [28] has failed to hold up under experimental scrutiny. Until such experimental refutation can take place, genetic theories of intellectual talent will contiue to hold sway among populationist behavioral scientists.

Instead, a number of writers have stooped to such devices as attributing guilt by association, aligning their populationist opponents with obvious crackpots of the past and attributing all manner of social and political biases to principal figures in the development of psychological tests. These are, of course, irrelevant; for example, replication of the principal experiments by workers in different parts of the world has long ago rendered certain data frauds, of which Cyril Burt was apparently guilty [29] of no importance.

For the moment, the paradigmatic split is likely to continue. The advance of biological thinking, with its attendant populationist leanings, into the behavioral sciences is likely to continue. Technical difficulties abound, and the scope of behavior is so great that the process is certain to take many years. The progress of science is always in danger of further obstruction from politically motivated interference, such as the infamous Lysenko affair in the Soviet Union, in which crackpot environmentalist theories in agriculture were officially imposed in place of sound biology. One can only hope that nothing like this lies in the future for behavioral science.

Scientists have one important lesson to learn from Kuhn. A little bit of self-consciousness goes a long way toward avoiding excesses. One's colleagues' morals and sanity are not put in doubt by their choices of working paradigm; only their scientific judgment is. Ultimately, experience will convince all of us who is more nearly right; in the meantime each of us must pursue the truth in his or her own way.

SEVEN

The New Science of Information

Howard L. Resnikoff

Civilization has been largely engaged in working out the social and economic consequences of energy-intensive machines that magnify muscle power since James Watt perfected the steam engine nearly two centuries ago. Today we are all participants—unwitting or unwilling though some may be—in working out the consequences of the new information-intensive machines that magnify mental power, whose social and economic consequences are likely to be even greater because these machines work on the stuff of the mind and bear upon those aspects of life that are distinctively human.

The theme of this essay is the new science of information. That there is a powerful interaction between science and technology is well known to students of either subject. In some cases science drives technology forward; in others, science is driven by advances in technology. Thus, modern astronomy emerged from the increased awareness of the heavens made possible by the invention of the telescope, and the microscope had an exhilarating effect on the development of the biological sciences, whereas general relativity and the mathematics that underlie the quantum theory owe little to the instruments of technology. It is, therefore, helpful to understand the relationship of a science and the corresponding technology.

The science of information is strongly technology-driven at the present time, for reasons that I hope to make clear, and information technology in turn is driven by economics. Before turning to the heart of our subject, let us briefly consider the interplay between the technology and the economic forces that provides the context in which the scientific issues are being addressed.

Information Technology and Its Societal Consequences

During the past century the composition of the American work force has changed dramatically. The fraction of the work force engaged in agriculture declined from more than 60 percent in 1860 to less than 5 percent in 1980. The fraction engaged in the production of goods peaked at about 35 percent in 1965 and has been precipitously declining since then. But the service sector of the economy, which is primarily concerned with information-related activities, broadly construed, has been growing rapidly. Similar changes are under way in industrialized nations throughout the world. This structural change in the work force is displacing people from energy-intensive and goods-intensive tasks into information-intensive tasks and thereby creates pressure for capitalizing information-intensive work and for developing a deeper understanding of the nature of information work and, consequently, of information itself.

Evidence is accumulating that labor productivity increases if workers are supported by adequate information technology, especially computing power. A recent study suggested that by providing a worker in an appropriate industry with one million AEG's (Active Element Groups; an AEG is a single bit of memory or a logic gate capable of one binary decision), productivity was increased by a factor of ten. Active Element Groups are being rapidly dispersed throughout society. A contemporary automobile incorporates about five microprocessors. The historical situation can be easily summarized: during the period of vacuum-tube electronic technology the annual rate of growth in the consumption of Active Element Groups in the United States was about 10 percent; with the introduction of transistor technology during the 1960s, annual increases in consumption grew to 24 percent; since the introduction of Large Scale Integrated Circuit technology, the rate of growth has jumped to about 105 percent per year.

At the present time the United States consumes enough Active Element Groups to equip every person on this planet with a pocket calculator. If present rates of growth of consumption of AEGs are sustained, by the year 2001 annual consumption will be great enough to equip each person with a computing machine having about as many Active Element Groups as the person has neurons in his or her brain.

To try to place this in perspective, compare the IBM 650 computer introduced in 1955 with the Texas Instruments TI59 programmable, hand-held calculator introduced in 1977. The former used 2,000 vacuum tubes, consumed almost 20,000 watts of power, weighed more than 5,500 pounds, required 10 tons of air conditioning, and cost $200,000 in 1955 dollars. This machine had a primary memory of 2,000 words stored on a magnetic drum, and it could add, multiply, and divide two numbers in 0.75, 20.0, and 0.5 milliseconds, respectively. The TI59 has 166,500 transistors on its integrated circuit chips, consumes 0.18 watts, weighs about two-thirds of a pound, and fits into a pocket. It has a 5,000-byte memory module and executes addition, multiplication, and division operations in 0.07, 4.0, and 0.4 milliseconds, respectively. It cost $299.95 in 1977 dollars.

Studies of semiconductor technology provide convincing evidence that a comparable advance relative to current technology will take place during the next two decades. Thus, the computing machine at the turn of the century will be to the TI59 as the TI59 is to the IBM 650. A qualitative improvement in information-processing capabilities of this magnitude will profoundly affect society, and it will also profoundly affect the conduct of science. The manufacturing processes currently used by the semiconductor industry routinely produce devices whose features are significantly smaller than many typical microorganisms, and laboratory research is carrying processing technology into the microworld, where the distinction between the animate and the inanimate becomes blurred.

These remarkable and unprecedented advances in information technology are raising profound questions for the science of information, but, more important, they are providing the experimental tools that are making it possible for the first time to investigate the inner properties of information and its governing principles that transcend the limits of direct human perception and analysis.

With these observations about the forces of economics and technology in mind, let us turn to the intellectual issues that underlie the scientific study of information.

The Science of Information

Information first began to assume an independent and quantifiable identity in the research of physicists and psychologists in the second half of the nineteenth century. It grew in importance with the development of electrical-based communication systems and has emerged as a unifying and coherent scientific concept along with the invention and development of computers during the past thirty years.

The subject matter of a science of information is inherently abstract, for it concerns not the substance and forces of the physical world but the arrangement of symbolic tokens or, as we may say, the structure of "patterns." Since symbolic tokens are necessarily represented by physical phenomena, as thoughts are represented by the electrical and biochemical states of the neural network, there is an intermixing of the physical sciences with the proper subject matter of information science. This makes it difficult to separate the pure properties of information science from those of the physical sciences. The latter merely describe the physical embodiment of the informational patterns.

It is clear that particular information can be presented in a variety of ways that differ inessentially from one another. For example, a propositional statement can be expressed in any one of the hundreds of natural languages in current use, or via symbols recorded on paper, or by acoustic speech wave forms. Evidently information is independent of the form in which it is presented. The proper generalization of this remark is that information is *that which remains after one abstracts from the material aspects of physical reality.* If one abstracts from the material reality of printed text, then all that is left is the organization of a sequence of symbols selected from a fixed but arbitrary finite inventory. In general, the concept of information is coextensive with the notion of order, or organization of material objects without regard to their physical constitution. Thus, the study of physically embodied information can be identified with the study of some type of relation of "equivalence" among classes of material configurations. Were it known how to axiomatize the properties of these equivalence classes, a formal "disembodied" theory of the structure of information would be the result.

In addition to studying the structure of patterns, a science of information must be concerned with the properties of information transfer. A relatively well-worked-out aspect of this problem falls under the heading of signal processing. Communication of information by real systems necessarily involves transfers of energy, so once again there is an unavoidable interplay between the physical substratum that carries the information and the patterned structures that constitute it.

Until recently the only known information-processing systems were biological. Today these systems have been joined by electronic information processors based on computers and telecommunications links that are, to be sure, still quite primitive in comparison with their biological cousins. The new science of information studies that higher level of abstraction that encompasses both the "bioware" and the "hardware" implementations—what men and machines have in common—rather than the details of either electronic or biological implementations of information-processing functions (which exemplify their differences).

Thus, this science seeks to uncover general measures and principles that may be expected to apply universally to information-processing systems and perhaps to the fundamental processes of nature themselves.

The patterned structures about which the most is known are those associated with the processing of sensory signals by biological organisms, especially by humankind. Neuroscience has uncovered much of value about bioware implementations. These limit and thereby partially define the means that govern the representation and processing of information. However, psychology and linguistics reveal our subject in a more nearly pure fashion. Thus, it is appropriate to consider the science of information from the standpoint of the cognitive sciences. But we should also recognize the limitations of psychology and linguistics as experimental sciences: although the initial conditions and externalities of the experimental arrangement can be modified by the experimenter, the structural properties of the system under investigation generally cannot be manipulated, nor can the system's internal state be prepared with any degree of certitude. This limitation does not necessarily deal a mortal blow to the role of experiment in the cognitive sciences: the paradigm of astronomy, most ancient and exact of the sciences, is limited to observation alone without control of even the initial conditions of the observed system. But the types of observations that can be made in the cognitive sciences are limited, and this limitation constrains the types of theories that can be readily tested by the experimenter. It is here that the role of the computer in information science is of crucial significance, for the computer scientist can prepare the internal state of the observed system as well as its initial conditions. For this reason the computer is the most important experimental tool in the history of information science, and it offers the promise of greatly accelerating progress in understanding the nature of information and the laws that govern it.

The history of information science consists of four interwoven strands: the study of thermodynamics and the theory of measurement in physics; the study of sensory information processing and knowledge representation in biological systems; the rise of electrical communication engineering; and the study of computability and the development of computing machines. In recent years the common features of these superficially disparate themes have begun to coalesce to form an integrated intellectual discipline whose principal problems blend elements of all four constituent fields.

Information and Thermodynamics

The recognition of information as a distinct entity in physical processes does not have a definite birthdate, but it appeared in an essential and ultimately highly developed form in the nineteenth-century study of thermodynamics. Thermodynamics has always occupied a curious position in physics. It is a phenomenological subject, and, despite revolutionary changes in the theories that underlie the composition and behavior of the fundamental constituents of matter and the material systems to which thermodynamics was applied, the accumulation of experimental evidence gathered over a long period of time has provided overwhelming support for thermodynamical laws. It is as if thermodynamical relationships have, ultimately, very little to do with the underlying structural theories of matter and their interactions. Thermodynamics is valid for Newtonian physical systems; it remains valid for systems governed by quantum mechanics; and it applies equally well and unchanged in the realm of relativistic phenomena. That this one part of physical theory could remain valid and essentially unaffected while the fundamental theories of physical composition and interaction underwent revolutionary gyrations suggests that perhaps thermodynamical descriptions are not, after all, descriptions of physical properties and relationships themselves but of some other, still more fundamental properties that are coeval with all regimes of physical inquiry and that can be abstracted from them. Were this true, thermodynamical relations might remain valid for all physical theories that describe an aspect of nature with an adequate degree of verisimilitude. This seems to be the case in the sense that thermodynamical quantities link physical entities to their organization, that is, to their informational properties, and that the informational structures are abstractions independent of the physical mechanisms that embody them. This is the viewpoint adopted by the science of information.

The celebrated Scottish physicist James Clerk Maxwell was the first to recognize the connection between thermodynamical quantities associated with a gas, such as temperature and entropy, and the statistical properties of its constituent molecules. There is no point here in trying to explain the meaning of the terms "entropy" and "temperature" because the original physical interpretations are rather complicated and not directly germane to our purposes, but also because the most illuminating interpretation turns out to be the informational rather than the physical one. Nevertheless, it was in this way that probability entered into thermodynamical considerations and therewith opened the road to further abstractions and generalizations of the entropy concept that were more broadly applicable to systems quite divorced in conception and properties from the gases and other fluids of physics. The measure of the entropy of a state (and hence, as was later realized, of its information content) in terms of probability was supplied by Ludwig Boltzmann's famous formula, $S = k \log W$, where S denotes entropy, W the number of accessible microstates, and k "Boltzmann's constant," which mediates between the physical units of entropy on the left side of the relation and the purely numerical quantity $\log W$ on the right; k is measured in units of energy per degree of temperature. Boltzmann himself recognized that entropy is a measure of disorder of the configuration of states of the atoms or other particles that make up a thermodynamical system, and hence it can also be considered as a measure of the order or degree of organization of the system. This viewpoint was clarified and sharpened by Leo Szilard, who in 1929 identified entropy with information and the measure of information with the negative of the measure of entropy, an identification that is the foundation of the quantitative part of the theory of information.

Szilard's contribution came about in the following way. An early hint that the statistical interpretation of thermodynamical properties must have something to do with information was already present in Maxwell's discussion in the mid-nineteenth century of an apparent thermodynamical paradox. He conceived two chambers separated by a common partition that can be removed in order to permit the objects in one of them to move freely into the other. If the two chambers are initially separated by the partition and one of them contains a gas (say, air under normal conditions of temperature and pressure), then, upon removal of the partition, the gas will rapidly diffuse to fill the entire chamber. But everyday experience tells us that the reverse sequence of events does not occur: the molecules of a gas distributed throughout a chamber will never congregate in one part so that, by slipping in a partition, the chamber can be divided into two parts, one of which contains all the gas and the other none of it. Maxwell proposed a "demon" who, he suggested, would be capable of performing this counter-intuitive feat. Here is Maxwell's idea. Suppose that the combined chamber is filled with gas and that the partition is reinserted. Further, suppose that the partition contains a small door just large enough to allow a molecule of gas to pass from one chamber to the other. This door, which can be opened and shut without frictional resistance, is attended by Maxwell's demon, who observes the gas molecules in one of the chambers and, when in its random motion one of the particles heads toward the door and is about to rebound from it, opens the door briefly, permitting the molecule to pass into the other chamber. Very soon there will be an excess of molecules in one of the chambers and a deficiency in the other that will, as time passes, become increasingly extreme, tending toward a final state in which the entire gas occupies but one of the chambers and the other is empty. Since the demon requires only an infinitesimal amount of energy to operate the door, this process appears to decrease entropy, in contradiction to the second law of thermodynamics.

This paradox was explained by Szilard in his 1929 paper, whose importance only became apparent much later. It was obvious, he pointed out, that in order to perform its task, Maxwell's demon had to be very well informed about the position and velocity of the molecules that approached the door so that it could judge when and for how long the door should be opened to enable the molecule to pass through and enter the chamber that would ultimately have an excess of the molecules without allowing any molecules to pass in the opposite direction. Szilard was able to calculate the entropy gained by the demon during the process of letting a molecule pass from one chamber to the other, and he showed that under the most favorable circumstances this entropy gain exactly balanced the entropy loss by letting the molecule pass through the door. Since the demon can play its role if and only if it has the necessary information, Szilard was led to identify *the negative of the demon's entropy increment* as the *measure of the quantity of information* it used. This identification of *information increments* with *entropy decrements* extended Boltzmann's interpretation of entropy as a measure of the orderliness (that is, the degree of organization) of a physical system and changed the focus of attention from the physical to the abstract aspects of the concept.

Early Development of Sensory Physiology and Psychophysics

While the physicists were busy interpreting the phenomenological notions of thermodynamics, first in terms of statistical mechanics and later in terms of more purely probabilistic and information-theoretic concepts, those scientists who studied biological information processing were not idle.

Ernst Weber, one of the founders of psychophysics, and his two scientist brothers had concentrated their research efforts on applying the exact methods of mathematical physics to the study of the various physiological systems of the higher animals and humankind. In 1826 Weber began a series of systematic investigations of the sensory functions that resulted in his formulation of the first general quantitative assertion of psychophysics. It can be stated as follows. Let the magnitude of some physical (that is, objective) stimulus be denoted by S and the magnitude of the corresponding psychological (that is, subjective) response be denoted by R. Weber found that if S is increased by an amount ΔS sufficient to create a just noticeable difference in response to the different stimuli levels, then the initial stimulus magnitude S and the stimulus increment ΔS are related by the equation $\Delta S/S =$ a constant, where the constant depends on the sensory modality and on the individual; for typical compressive modalities (for example, perception of loudness or of brightness) the constant is approximately equal to $\frac{1}{30}$. It is known today that the "Weber fraction" $\Delta S/S$ is approximately constant for only a limited range of sensory stimuli, but Weber's observation was a key and fruitful result that itself was the stimulus for much later work.

In 1850 Theodore Fechner, basing his analysis on these earlier investigations of Weber, conceived the psychophysical function, which expresses the subjective measure of magnitude of the psychological response in terms of the objective measure of magnitude of the physical stimulus, and proposed that it could be represented by a logarithmic dependence of response magnitude on stimulus magnitude. In terms of the stimulus magnitude S and the response magnitude R, this means that the psychophysical function $R = f(S)$ is given, according to Fechner, by the formula $R = a\log S$, where a is a constant that characterizes the sensory modality and the individual, and $\frac{1}{a}$ can be identified with the Weber fraction. A function of this type compresses a great range of variation of the stimulus magnitude into a relatively small range of variation of the response magnitude; thus, the latter is more readily assimilated by the biological information-processing gear that is, after all, made from unreliable components that are slow, bulky, and require special temperature and other environmental conditions in order to function.

Fechner's result has been frequently challenged, perhaps most effectively by S. S. Stevens; indeed, it is obvious that the logarithm alone cannot account for the qualitative behavior of sensory response to either very large stimuli or, without suitable modifications for a "threshold" level of activity, to very small stimuli. In the former case, $\log S$ increases without bound as S does, but real sensory-processing channels have a limited capacity. Nevertheless, recent studies have shown that the neural-firing frequencies that encode sensory stimuli do appear to be proportional to the logarithm of the stimulus magnitude throughout much of the normal range of stimuli experienced by an organism. This provides a heretofore absent level of detailed confirmation of the essential validity of both Weber's and Fechner's pioneering work.

Hermann von Helmholtz was the greatest figure in nineteenth-century physiology and psychophysics, and one of the greatest scientists in general. He was interested in everything and made fundamental contributions to many aspects of physics, mathematics, and biology. It was Helmholtz who formulated the law of conservation of energy in what was essentially its modern form.

In 1850, the same fruitful year when Fechner conceived the psychophysical function, Helmholtz made the first measurement of the velocity of propagation of neural impulses. Earlier physiologists and physicists had held that the velocity of neural signals was so great as never to be measurable by experiment within the compass of an animal body. In fact, electrical impulses propagate along neurons at speeds that vary between about 10 meters per second and 100 meters per second, faster propagation corresponding to nerve cells whose axons are myelinated and have the greater diameter; the faster neurons have a channel capacity on the order of several thousand bits per second. Just as the discovery that light is propagated with a large but finite speed influenced theoreticians two centuries earlier, the measurement of the speed of neural impulses profoundly conditioned and limited the kinds of theories of sensory and mental activity that would thereafter be entertained.

Helmholtz's results made it evident that not all information incident upon the sensory system could be processed by the neural net in "real time," that is, rapidly enough for the organism to respond to sensed changes in its environment. If the speed of propagation were infinite, then an infinite amount of computation could be performed by a neural network in any brief interval, so there would be no computational barrier to the full processing of all sensory input signals. If the speed of propagation were finite but very great, say, comparable with the speed of light, there would be theoretical limitations on computational capabilities and the time required to process complex signals but they would not impose significant practical constraints. Neither of these two possibilities conditions the computational models of biological information-processing systems, and neither has much explanatory power to help the scientist interpret experimental data or discriminate among alternative theories.

The truth of the matter is that neuron-based biological information processing is very slow. The speed of propagation of neural signals implies that it must take between 0.01 second and 0.1 second for a signal sent by the brain to reach the hand or foot, or as long as 3 seconds for one sensed by the tail of the dinosaur *Diplodocus* to reach its head. It is the low speed of neural signal propagation that makes the illusion of continous motion possible in motion pictures that successively display 24 still frames per second. This means, in particular, that the human vision system cannot distinguish continous physical motion from a discontinous succession of images if the rate of image presentation is greater than some small number per second. If neural signals propagated at infinite speed, we would have no difficulty in detecting discontinous motion at any rate of presentation whatever, and the illusions so admirably created by television and cinema photography would be impossible. Helmholtz observed that

> Happily, the distances our sense-perceptions have to traverse before they reach the brain are short, otherwise our consciousness would always lag far behind the present, and even behind our perceptions of sound.

Taking the small finite value of the rate of transmission of neural impulses into account, an elementary estimate of the quantity of information that is, for example, incident upon the retina of the human eye shows that it greatly exceeds the transmission capacity of the neurons that lead from it to the higher cognitive centers. A similar assertion can be made about the capacity of the ear compared with that of the channels that conduct aural information to the brain. Most of the incident information must be omitted by the higher processing centers.

Several years after the discoveries of Helmholtz and Fechner, the mathematician Hermann Grassmann showed in 1853 that the space of physical light signals of all visible intensities and frequencies, which is an infinite-dimensional *objective* space, is compressed by the human vision system into a region in an abstract three-dimensional *subjective* space of color perception, the so-called "color cone."

Taken together, the results of Helmholtz, Fechner, and Grassmann suggest that sensory data is corrupted and modified by the sensing organism in ways that greatly reduce its effective quantity and substantially modify its original form.

Selective Omission of Information

These corruptions and compressions of sensory data have the effect of reducing an unmanageable glut of input information to an amount that can be processed by the mental computing equipment fast enough to be useful for responding to changing environmental circumstances. There are many different ways the incident information could be reduced in quantity. The receptive system could, for instance, filter the input to uniformly limit the amount of data received through each input channel. This would preserve some fixed fraction of the information in the original signal without giving preference to any particular kind, but, because the total amount of information sent forward to the higher processing centers would be reduced, the reduction would be noticed more in those informational constituents that are the more "essential" for the organism's purposes. Were it possible to omit information selectively, retaining the "essential," or at least the "more essential," portions while rejecting the "less essential" ones, then the computational load on the mental computer could be substantially reduced, and its ability to respond to sensory inputs in a timely way would be correspondingly increased. This is just what all biological information-processing systems do. But how the discrimination between "more essential" and "less essential" is made is not yet fully understood, although certain basic features of the process are reasonably clear. This discrimination, which occurs to a considerable degree in the peripheral elements of the neural net adjacent to and connected to the sensory organs themselves, embodies a critically important pattern-classification mechanism that operates prior to the activities of the higher processing centers and limits the information that is available to the latter. It would be a particularly valuable contribution to understand the principles governing this preliminary pattern classification in which a fundamental discrimination is created between the part of the signal carrying information of general use to the organism and the remainder, which is of equal physical significance but can be treated like "noise" from its informational standpoint.

The need to selectively omit information occurs in internal mental processing as well as in the processing of sensory stimuli. Consider, for example, the mental analysis of a chess position. The game tree, which determines all possible consequences of alternative plays, has a combinatorial complexity that far surpasses the ability of the human brain or any computing machine to explore it completely; it is estimated that there are as many as 10^{120} different possible games. Humans reduce the complexity of this search problem by selectively omitting most of the pathways. The omitted ones are expected to be unproductive, but the process by which a person makes the decision to omit particular paths remains unknown. The computer scientist usually calls selective omission in the context of a decision problem that involves searching through a complex tree the application of "heuristic" procedures. But this is just a term to neatly name an essential process that is still largely a mystery.

It is important for us to realize that selective omission occurs not only at the level of omitted data, but also—perhaps more importantly—at the level of omitted logical procedures, procedures that could be applied to data but, as the result of a systematic process of exclusion, are not.

The Quantum Mechanical Theory of Measurement

Apart from the general argument that macroscopic information process-ing and measurement constraints are ultimately consequences of the limitations that exist at the most fundamental level of physical phe-nomena, many of the direct macroscopic measurements, made either by means of laboratory and other instruments or by the physiological de-tectors of sensation, operate with the same materials and means that support the foundations of physical measurement. For example, the eye is sensitive to photons of light that are governed by the wave-particle duality of quantum physics at the level of their interaction with the photosensitive chemical receptors in the rods and cones of the retina. Therefore, it will be appropriate for us to summarize the development of certain aspects of the theory of measurement from the viewpoint of quantum physics in order that we may be able to form an impression of the nature of the limitations that appear to govern information process-ing at the most basic physical level.

The breakdown of classical physics, which became increasingly ap-parent at the end of the nineteenth century, was due to the improper ex-tension of naive concepts of measurement and measurability from the realm of everyday life to circumstances increasingly distant from their origins. For our purposes, one of the central new developments that emerged in the the first thirty years of the twentieth century was the recognition of the interdependence of measurements of variables of dif-ferent kinds. In striking contradiction to ideas that had been accepted as the basis for science for more than two hundred years, it was discovered, for example, that the accuracy of measurement of the *position* of an ob-ject interfered with the accuracy of measurement of its *velocity* of mo-tion (more precisely, of its *momentum*), and that intervals of time and space have only a relative meaning, depending on the motion of the ob-server. Thus, there is an inherent ambiguity, an uncertainty, in the pos-sible knowledge that we can have of our physical surroundings. This re-striction has important practical as well as philosophical consequences, and it affects the stability and structure of chemical combinations in a direct way. Erwin Schroedinger, one of the creators of quantum mechan-ics, saw in this property (coupled to certain other aspects of the new physics) the secret to the existence of life itself: that is, why life is pos-sible amid the incessant battering of the genetic material in a sea of thermal agitation.

That energy was bundled in discrete units, "quanta," was discovered by Max Planck in 1900 in the course of his investigations into the inade-quacies of the classical desciption of the thermodynamical properties of radiation confined in an enclosure. This key discovery led, through a complex and indirect pathway, to the creation of a coherent new fundamental theory of physical phenomena by Werner Heisenberg in 1925. It was Heisenberg who discovered that the accuracy of measurement of "conjugate" physical variables was mutually constrained by the inequalities of uncertainty that bear his name, and that all physical measurements are, ultimately, statistical in the sense that repeated measurements of identical physical situations will not lead to identical results but only to measurements that are collectively related to one another by a statistical law.

These discoveries seem to place the basic structure of physical nature on the same slippery footing as human affairs: variable subjective responses are the consequences of fixed objective stimuli; accurate statistical descriptions of, for example, the distribution of longevity for a collection of people are easily come by, but it is not possible to predict the life span of an individual.

The deep questions about the nature and limitations of measurement were taken up by the polymath John von Neumann, who in 1932 attempted to formalize and axiomatize the new physical theory. His pathbreaking book included an extensive analysis of the problem of measurement and showed how it was related to the thermodynamical and informational notions that had independently grown from the considerations of physicists such as Boltzmann and Szilard. In effect, von Neumann showed that the observer must be taken into account in the process of measurement but that the dividing line between observer and observed is fluid. If the combined system, consisting of the observed portion and the observing portion, is treated according to quantum mechanics, then the total amount of information in the course of a measurement is preserved. But at some ultimate stage the psychological "external" consciousness that assimilates the observation leads to an irreversible increase in entropy. This in an unusually subtle matter. It means that information gained about some variable or event must be compensated by a precisely equal amount of information lost about some other variable or process and, possibly, by some additional irreversible information loss. Although at the most fundamental level of measurement theory such a "law" of conservation of information may hold, at the macroscopic level it is not possible for us to trace all the information flows that result from a measurement. The consequences are that some information "gets lost" whenever a measurement is performed and that the degree of organization of a static universe continually decreases.

At the cosmic scale this result seems to contradict both observation and intuition, for we see around us a highly organized universe. It is not enough to argue, as most scholars have done in the past, that the high degree of organization we observe is a "local fluctuation" that must be paid for by a greater than average degree of disorganization elsewhere in the universe.

Recently the idea has been offered that the expansion of the universe and the force of gravity act in combination as an organizing principle by increasing the number of states of the physical universe more rapidly than its material content can change to occupy them. This ultimately acts to *decrease* entropy and *increase* organization. But much work remains to be done before this argument can be said to be compelling.

Early Development of Electrical Communication Engineering and Its Formalization as Information Theory

The third of the historical threads that combine to form the science of information is the development of electrical communication engineering. Communication engineering refers to communication by means of electromagnetic signals, including the transmission of electrical currents through wires, as in telegraphy and telephony, and the transmission of electromagnetic waves through space, as in broadcast radio or television. The subject grew from the practical needs of engineers to understand how to design complex systems for efficient performance. Because variation in electric currents and electromagnetic fields cannot be directly sensed by people, except for the small band of radiant electromagnetic energy called "light," the methods used in this practical field are more theoretical and more dependent on measuring instruments than are those of other engineering disciplines. Indeed, the boundary between "science" and "engineering" is often scarcely definable. As a result, the interplay between theory and practice has been unusually intense and has stimulated the exceptionally rapid development of the subject.

As it turns out, the general principles that govern the transmission of information by electrical means extend to other media as well; moreover, the discovery that the neural network transmits information by encoding it as chemically generated electrical impulses suggests that the tools built by the communications engineer should be of great value in the general study of neurologically mediated information communication. This was first clearly anticipated by Norbert Wiener and has indeed turned out to be the case.

The most decisive contribution of communication engineering was the discovery of the proper measure of information in the context of the transmission of information. Although arrived at from an entirely different standpoint, the result coincided with what the statistical thermodynamicists had found: the measure of information is formally identical to the negative of the entropy measure, assuming, of course, an appropriated interpretation of the corresponding concepts. Let us recall that a communication system functions by successively transmitting signals drawn from a certain inventory, such as the collection of letters of the alphabet. If the kth signal in the inventory is transmitted with a probability p_k, then the measure of information provided by the transmission of a single signal selected from the inventory is $I = -\Sigma p_k \log_2 p_k$, where the summation runs over all signals. R. V. L. Hartley of the Bell Telephone Laboratories offered a logarithmic definition of the quantity of information in 1928, but it did not take into account the case of unequal probabilities, and, as he was unable to erect a predictive theory that made significant use of the definition, it languished until twenty years later, when the deficiencies were remedied by Claude Shannon. But Hartley understood the engineering issue at stake precisely. In the introduction to his paper he described his goal by stating:

> What I hope to accomplish . . . is to set up a quantitative measure whereby the capacity of various systems to transmit information may be compared.

Other researchers were thinking about similar problems. In 1922 the distinguished statistician Ronald Fisher had attempted to define the information content of a statistical distribution and had suggested the reciprocal of the variance as a candidate. But World War II provided the impetus for bringing scientists and engineers together to focus their intellectual energies on the problems of communication in the context of the development of radar, and that may have provided the main stimulus to the creation of an essentially complete theory of communication by Claude Shannon. Shannon was a student of Norbert Wiener, who appears to have independently suggested the measure of information that today is usually associated with Shannon's name. But it was the latter who used this measure to create a theory that could be applied to a wealth of problems of the most general variety to determine the efficiency of a communication channel and to aid in the design of communication systems. Shannon's pioneering work was published in a series of three papers in 1948. Although his work has been the subject of innumerable investigations since then, within its framework no fundamental new idea or result has been added to the elegant edifice he constructed.

The Development of Computers

The whole of the developments and operations of analysis are capable of being executed by machinery.—Charles Babbage

The fourth and final strand in the history of information science is the invention and development of the stored-program digital computer. The idea of a special-purpose machine that could compute and perform certain restricted information-processing tasks is very old. In antiquity, *analog* devices were used to aid astronomical calculations. An ancient calendar computer from around 80 B.C. has been recovered from the wreck of a Greek trading vessel off Antikythera. The abacus, another ancient device, is an example of a *digital* aid to computation. The concept of a machine that could perform an unrestricted range of mathematical calculations dates at least to the time of the Thirty Years War, for in 1623 Wilhelm Schickard, a professor of astronomy, mathematics, and Hebrew at Tübingen, designed and built a mechanical device that performed addition and subtraction completely automatically and partially automated multiplication and division. This machine appears to have had no discernible effect on his contemporaries. A generation later, in 1642–44, Blaise Pascal, then twenty years old, invented and built a simple machine that performed addition and subtraction. This device seems to have been inferior to Schickard's, but it nevertheless received considerable attention and was later described in detail in the famous *Encyclopedie* of Diderot.

Leibniz appears to have been the first to conceive of a universal logical machine that would operate by a "general method in which all truths of reason would be reduced to a kind of calculation." The earliest attempt to put this idea into practice was made by Charles Babbage, who not later than 1822 was already toying with the notion of a steam-driven machine that could be applied to the onerous general calculations that arise in astronomy. The British government subsidized Babbage's work for twenty years, but his machines were never completed. Although the ideas behind them were correct, complex computing machines cannot be built from mechanical parts because of the cumulative effects of wear and the inherent inaccuracy of their relative positions, both limitations being more fundamental than they may at first appear.

The principal theoretical impetus to the development of the general-purpose digital computer can be traced back to the last decades of the nineteenth century. While scientists were struggling with the newly uncovered inadequacies of classical physics, logicians and mathematicians were becoming correspondingly uneasy about paradoxes and contradictions that were emerging from the logical processes of reason and the foundations of mathematics. As a result of attempts to arithmetize mathematical analysis, a general theory of sets was created by Georg Cantor between 1874 and 1897, but it was soon cast into doubt by the discovery of numerous paradoxes, of which Bertrand Russell's is perhaps the most widely known. The Russell paradox, discovered in 1902, concerns the set of all sets that are not members of themselves. Is this set a member of itself? Straightforward application of the rules for reasoning show that it is; but then, according to its definition, it cannot be. Paradoxes of this kind raise the question of the definability of mathematical objects. At about the same time, mathematicians became concerned about the *consistency* of mathematical systems such as classical number theory and elementary geometry—whether the application of logical processes of proof to a collection of axioms might ever result in contradictory statements—and about their *completeness*—whether all true statements about, say, elementary geometry, could be logically derived from a collection of consistent axioms. These and related questions exercised mathematicians such as David Hilbert and John von Neumann as well as logicians and philosophers, who, in order to cope with the increasingly technical nature of the problems, found themselves increasingly adopting mathematical formulations and tools.

In 1925 Werner Heisenberg discovered the physical limitations to our ability to observe natural phenomena that are expressed by the uncertainty principle. A different kind of limitation and uncertainty principle for the processes of reason was revealed by the mathematician Kurt Gödel in 1931. He showed that any mathematical system whose structure is rich enough to express the ordinary arithmetic of integers is necessarily incomplete: there will be true statements that cannot be proved within the system. This lack cannot be made good by incorporating such an unprovable true statement as an axiom in the system, for then other unprovable truths emerge. This is another way of saying that there are inherently undecidable statements whose truth cannot be established by merely drawing out the logical consequences of any collection of axioms.

Gödel's method, as well as his result, was important because it established a link between arithmetic and logic by coding the propositions and logical operations in arithmetic terms, thereby effectively reducing questions about logic to questions about computation. The complete identification of logical reasoning with computation was proposed by Alonzo Church in 1936 and was implicitly adopted in 1936–37 by Alan Turing, whose conception of a universal calculating machine (the "Turing machine") marks the birth of the modern theory of the digital computer. This abstract machine is able, as Turing proved, to compute anything that can be computed by any special (deterministic) machine; in this sense it is "universal" and, consequently, a valuable model for theoretical analysis.

The course from a fledgling theory of computability to the deployment of practical and powerful computing machines has been run with remarkable rapidity. Spurred on by the needs and means provided by World War II, researchers made great strides in the design of computing machines and in the underlying electronic technology needed to implement those designs. By 1945 the key concept of the stored-program computer had been created by the same John von Neumann who a decade earlier had been analyzing the nature of observation and measurement in quantum mechanics and who would a decade later concentrate his mental powers on investigations of the working of the brain and of self-reproducing automata—that is, of life.

It did not take very long for the concept of the Turing machine and its partial realization by practical and increasingly powerful general-purpose logical (computing) machines to lead to the ideas that the brain is perhaps nothing more than a spectacularly ingenious computer built from error-prone components wired together in a highly variable way, and that the continuity of existence of living species from one generation to the next is maintained by a kind of "program" coded in the macromolecular proteins of the genetic material. These are special instances of the more general idea that all of the features peculiar to living organisms and to life itself can be understood in terms of information-processing principles inherent in the computer concept. Freeman Dyson recently expressed this view when he wrote,

> . . . as we understand more about biology, we shall find the distinction between electronic and biological technology becoming increasingly blurred.

The Information-Processing Viewpoint in Psychology

Both the computational interpretation of the process of thought and the signal-processing viewpoint of the communications engineer spread rapidly, even before computers having significant capabilities had been constructed. Within five years of the publication of Shannon's basic papers and the construction of the first stored-program computer, many investigators had begun to explore the implications of both developments for psychology.

As early as 1947 Walter Pitts and Warren McCulloch published a seminal paper with the provocative title, "How We Know Universals: The

Perception of Auditory and Visual Forms," in which they proposed a mechanism, based on ideas from the mathematical group theory, by which the neural net extracts universals from sensed data. Very much in the spirit of the signal-processing and the "brain as a computing machine" viewpoints, their work established the path that many others would later follow and that continues to influence research today.

The next year, 1948, saw the publication of the first edition of Norbert Wiener's influential and remarkably prescient *Cybernetics: or Control and Communication in the Animal and the Machine*, which identified the important role of feedback in governing the activities of machines and animals and which adopted for the first time a completely comprehensive view of animal performance and capabilities from the computing and communications-engineering standpoints. The discussion of how information can be quantified already appears here from the standpoint best suited for general application to the measurement process.

By 1950 the question of whether the brain is merely a bioware implementation of a computing machine had become specialized into a spectrum of particular issues, with "whether" giving way to a more experimentally oriented "how": how could a machine be programmed to solve general problems, later addressed by Allen Newell, J. C. Shaw, and Herbert Simon, and how could the behavior of a suitably powerful and properly programmed machine be distinguished from the behavior of a person?

The latter question directed attention to the problem of deciding whether the abilities of computing machines were for all practical purposes indistinguishable from those of humans, which is different from the theoretical question of whether the brain is in fact merely a computing machine. Turing took up the matter and in 1950 proposed his famous "test." In 1951 Marvin Minsky produced an early analog simulation of a self-reproducing system. In 1953 Loren Riggs and his colleagues, in their investigations of human vision, made the remarkable discovery that images stabilized on the retina vanish; this has both signal-processing and higher categorical implications for the way the brain processes sensory information to produce the universals for which Pitts and McCulloch had been searching. There were two notable developments in 1954. Fred Attneave showed that the information content of two-dimensional images was nonuniformly distributed, and he identified some of the preferred features in which it resides. His famous "cat" demonstrated that the major part of the information contained in a contour is concentrated near points of extreme curvature. In the same year George Miller published a fundamental paper that described the short-term memory information-processing capacity of the human brain. The capacity is surprisingly limited: as the catchy title of his paper expressed it, only

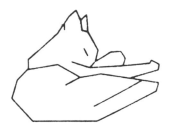

Attneave's "Cat"

"seven plus or minus two" chunks of information can be held in short-term memory. This is not the place to make the meaning of the term "chunk" precise; suffice it to say that if the input items are a sequence of bits, then about seven of them can be retained in short-term memory, but that the same statement could be made with equal validity for other, larger units, such as ordinary decimal digits or words drawn from the vocabulary of a natural language. Miller's result tells us why local telephone numbers consist of seven digits, and it also tells the post office why there is considerable resistance to eleven-digit zip codes.

In 1958 von Neumann's book, *The Computer and the Brain,* appeared, and in 1959 Leon Brillouin published his ambitious attempt to incorporate all of scientific endeavor within the formal precincts of Shannon's information theory. Although Brillouin was not successful in providing a completely coherent theoretical structure that is capable of predicting new results as well as explaining old, his viewpoint acted as a stimulant and a signpost, hinting at the kind of theory that might ultimately be created. In the same year, Lettvin, Maturana, McCulloch, and Pitts analyzed the signal-processing capabilities of the frog's eye and were able to determine what it tells the frog's brain. Not much, as it turns out, although the frog gets along very well, thank you, and therein lies the great significance of this important paper for the development of a realistic theory of knowledge representation.

In 1961 a corrected and enlarged second edition of Wiener's *Cybernetics* appeared. In the preface to the new edition the author expressed his opinion that, whereas in 1948 cybernetics was "merely some program to be carried out in the future," by 1961 it was time to reconsider it "as an existing science." Seventeen years later James Grier Miller's effort to create a general theory of *Living Systems* (the title of his book) showed how deeply the ideas of information processing had permeated the scientific community and the extent to which Wiener's program had indeed become "existing science." In recent years evidence about how knowledge is represented in animal brains and on the pattern-classification capabilities of organisms has steadily accumulated. Concerning human abilities, attention has been focused on the use of visual illusions as an experimental tool for investigating the aggregated nature of image processing. In addition to the traditional illusions known for many years, new ones based on the study of stabilized retinal patterns and on subjective contours provide interesting insights into the information-processing mechanisms that the brain exploits to reduce the quantities of data it must process in order to sustain the illusion of reality. Recent work of Roger Shepard, of Stephen Kosslyn, and of Steven Pinker has identified some of the key features of the mental storage and manipulation of three-dimensional image information and has provided evidence about the computational mechanisms they involve. Francis Crick, David Marr, and Tomaso Poggio have coupled sophisticated results from communication theory to neurophysiological investigations and produced a striking information-processing explanation of visual acuity.

I do not hesitate to predict that we stand at the beginning of discoveries that will firmly link our understanding of mechanical and biological information processing by uncovering general principles of pattern structure that apply throughout nature. I will even venture so far as to say that theories of the nature of the physical universe will, ultimately, be reformulated in terms of the twin primitive notions of energy and information, whose intimate combinations we see reflected in the world about us.

NOTES

TWO

1. "The Letter to Can Grande" (*Epistola X ad Canem Grandem della Scala*), in *The Literary Criticism of Dante Alighieri*, ed. and trans. Robert S. Haller (Lincoln: Univ. of Nebraska Press, 1973), p. 99.

2. See, for example, *Iliad* 9.443 and 19.443 and *Odyssey* 11.561.

3. See *Odyssey* 3.94.

4. Northrop Frye, in "New Directions from Old," *Myth and Mythmaking*, ed. Henry A. Murray (New York: George Braziller, 1960), p. 117.

5. Roger Hinks, *Myth and Allegory in Ancient Art*, Studies of the Warburg Institute (London: Warburg Institute, 1939), p. 4.

6. For the use of allegory's four senses in scriptural exegesis and also in sermons, see Harry Caplan, "The Four Senses of Scriptural Interpretation and the Mediaeval Theory of Preaching," *Speculum*, 4 (1929), 282–90. For the use of Neoplatonic cosmology or pagan myth in *involucrum*, see M. D. Chenu, "*Involucrum*: Le mythe selon des theologiens medievaux," *Archives d'histoire doctrinale littéraire du Moyen Âge*, 30 (1955), 75–79. For the distinction between the two, see also Winthrop Wetherbee, *Platonism and Poetry in the Twelfth Century: The Literary Influence of the School of Chartres* (Princeton, N.J.: Princeton Univ. Press, 1972), pp. 36–48; and especially Hennig Brinkmann, "Verhullung ('Integumentum') als literarische Darstellungsform im Mittelalter," in *Der Begriff des Repraesentatio im Mittelalter: Stellvertretung, Symbol, Zeichen, Bild*, in *Miscellanea Mediaevalia*, vol. 8 (Berlin and New York: Walter de Gruyter, 1971), pp. 314–39.

7. "Integumentum est genus demonstrationis sub fabulosa narratione veritatis involvens intellectum, unde etiam dicitur involucrum." See the recent edition by Julian Ward Jones and Elizabeth Frances Jones, *The Commentary on the First Six Books of the Aeneid Commonly Attributed to Bernardus Silvestris* (Lincoln and London: Univ. of Nebraska Press, 1977), p. 3; and also *Commentary on the First Six Books of Virgil's Aeneid by Bernardus Silvester*, trans. Earl G. Schreiber and Thomas E. Maresca (Lincoln and London: Univ. of Nebraska Press, 1979), p. 5.

8. *Policraticus* 8.24, ed. C. C. J. Webb, vol. 2 (Oxford: Oxford Univ. Press, 1909).

9. D. W. Robertson, Jr., "Some Medieval Literary Terminology, with Special Reference to Chrétien de Troyes," *Studies in Philology*, 48 (1951), 669–92, esp. 692; see also his "Marie de France, *Lais, Prologue, 13–16*," *Modern Language Notes*, 64 (1949), 336–38, for discussions of *littera, sensus*, and *sententia*. Brinkmann, pp. 314–39, discusses the following terms: *integumentum, involucrum, aenigma, allegoria, ironia, antiphrasis, euphemismus, sarkasmus, icon, parabola, paradigma, parable, metonymy, prosopopoeia, apostrophe, conformatio, mutatio, transmutatio*, and *symbolum*, many of which are rhetorical rather than grammatical terms.

10. J. Reginald O'Donnell, "The Sources and Meaning of Bernard Silvester's Commentary on the *Aeneid*," *Mediaeval Studies*, 24 (1962), 233–36.

11. Diogenes Laertius, 8.21 of *Lives of the Eminent Philosophers*, trans. R. D. Hicks, 2 vols. (London: William Heinemann; New York: Putnam, 1925), II, 339. This volume contains information about the pre-Socratic and Stoic philosophers and the Platonists.

12. Fragment 11 of Hermann Diels, *Die Fragmente der Vorsokratiker*, 3 vols. (Berlin: Weidmannsche Verlagsbuchhandlung, 1952; rpt. 1960), English trans. by Kathleen Freeman, in *Ancilla to the Pre-Socratic Philosophers: A Complete Translation of the Fragments in Diels, Fragmente der Vorsokratiker* (Oxford: Basil Blackwell, 1952).

13. Fragment 42, Diels; Freeman, p. 28.

14. Fragment 56, Diels; Freeman, p. 28. Basically Heraclitus did not believe in the gods and discounted the poets who used them: "What intelligence or understanding have they? They believe the people's bards, and use as their teacher the populace, not knowing that 'the majority are bad, and the good are few,'" Fragment 104, Diels; Freeman, p. 31.

15. Fragment 5, Diels; Freeman, p. 14.

16. Origen, *Contra Celsum*, 6.43, trans. Henry Chadwick (Cambridge: Cambridge Univ. Press, 1965), p. 359. See also J. Tate, "On the History of Allegorism," *The Classical Quarterly*, 2 (1934), 107, on the role of Pherecydes.

17. For the Porphyry *scholium* citing Theagenes, see the *Scholium Venetus B on the Iliad* 20.67 in William Dindorf, ed., *Scholia Graeca in Homeri Iliadem*, vol. 4 (Oxford: Clarendon Press, 1877), p. 231.

18. Mentioned by Diogenes Laertius, 2.11.

19. Cited by the Byzantine Syncellus, *Chronika* 140C.I, in Fragment 61, no. 6, Diels.

20. According to John Edwin Sandys, *A History of Classical Scholarship*, Vol. 1: *From the Sixth Century B.C. to the End of the Middle Ages*, 2d ed. (Cambridge: Cambridge Univ. Press, 1906), p. 30; see also Tzetzes on the *Iliad*, p. 94, ed. God. Hermann.

21. Fragment 19, Diels; Freeman, p. 86.

22. According to Tatian's comments on the *De Homero* of Metrodorus in *Oration adversus Graecos*, in *Patrologiae Cursus Completus: Series Graeca*, ed. J. P. Migne (Paris, 1844–64), 6, 854. He regards Metrodorus as childish in his treatment of the Homeric gods.

23. These identifications of Metrodorus come from a variety of sources: Hesychius, who equated Agamemnon with aether, and the Herculanean Rolls, authored supposedly by Philodemus (according to Th. Gomperz in his article on Metrodorus in *Sitzungsberichte der Wiener Akademie*, Phil. hist. classe, 116 [1873], 12–14), which equated the other gods and heroes with the appropriate natural forces, cited in Diels 61, no. 4. See also the discussion in Félix Buffiere, *Les Mythes d'Homère et la pensée grecque* (Paris: Société d'Édition 'Les Belles Lettres,' 1956), p. 127.

24. Fragment 30, Diels; Freeman, p. 8: "Of the reasoning men, a few, raising their hands thither to what we Greeks call the Air nowadays, said: Zeus consid-

ers all things and he knows all and gives and takes away all and is King of all."

25. Fragments 1b and 2, Diels; Freeman, p. 91.

26. Cited by Sextus Mathematicus 9.18, in *Opera*, vol. 3, ed. Hermann Mutschmann, rev. J. Mau (Leipzig: B. G. Teubner, 1961).

27. Fragment 52a and 53, Diels; Freeman, p. 39.

28. See Plato's discussion of names and their relation to reality in *Charmides* 163D in *Plato*, vol. 8, trans. W. R. M. Lamb (London: William Heinemann; New York: Putnam, 1927), and in *Cratylus* 391D–397E in *Plato*, vol. 6, trans. H. N. Fowler (London: William Heinemann; New York: Putnam, 1926). See also Cicero, *De natura deorum* 1.118.

29. Ann Bates Hersman, *Studies in Greek Allegorical Interpretation* (Chicago: Blue Sky Press, 1906), pp. 17–18.

30. Zeno of Citium wrote, according to a list of his works supplied by Diogenes Laertius 7.4, and Cicero, *De natura deorum* 1.15, five books entitled *On Homeric Problems* and also the scholia to Hesiod's *Theogony*, on verses 134 and 139. All the Stoics, according to Cicero, wrote to show "physica ratio non inelegans inclusa est in inpias fabulas." See Cicero, *De natura deorum* 2.24, in *De natura deorum; Academica*, ed. and trans. H. Rackham (London: William Heinemann; New York: Putnam, 1933). All citations from the Latin text derive from this edition. Zeno of Citium rationalized the gods, especially Juno and Jupiter, by "arguing that these were merely names given symbolically to mute and inanimate forces," in Cicero, *De natura deorum* 1.36.

31. According to Minucius Felix, *Octavius* 19.10, in *Tertullian; Minucius Felix*, trans. Gerald H. Rendall, based on the unfinished version by W. C. A. Kerr (London: William Heinemann; New York: Putnam, 1931).

32. Cicero, *De natura deorum* 1.15, and Diogenes Laertius, 7.135.

33. Cicero, *De natura deorum* 3.24.

34. Stanley Tate Collins, *The Interpretation of Vergil, with Special Reference to Macrobius* (Oxford: B. H. Blackwell; London: Simpkin, Marshall, 1909), p. 12. A list of Cleanthes' works appears in Diogenes Laertius, 7.175. See also the discussion of the followers of Zeno in Cicero, *De natura deorum* 1.37, 39–40, 41.

35. Cicero's title resembles titles of works by Zenocrates and Chrysippus. See J. M. Ross, Introduction, in Cicero, *The Nature of the Gods*, trans. Horace C. P. McGregor (Middlesex: Penguin, 1972), p. 18.

36. See Cornutus, *Theologiae graecae compendium*, ed. C. Lang (Leipzig, 1881); and Heraclitus of Pontus, *Allégories d'Homère*, ed. and trans. Félix Buffière (Paris: Société d'Édition 'Les Belles Lettres,' 1962).

37. Writers like these included Justin Martyr, Minucius Felix, Tertullian, and Clement and Origen of Alexandria. See Henry Osborn Taylor, *The Emergence of Christian Culture in the West: The Classical Heritage of the Middle Ages*. 3d ed. 1911; rpt. (New York: Harper, 1958) pp. 12, 14.

38. See Don Cameron Allen's interesting first chapter in *Mysteriously Meant: The Rediscovery of Pagan Symbolism and Allegorical Interpretation in the Renaissance* (Baltimore and London: Johns Hopkins Univ. Press, 1970), esp. pp. 3–18. Among the idolators attacking Christianity whom he mentions are

Valerius Maximus, Horace, the younger Pliny, Tacitus, Suetonius, and, later, Lucian or Philostratus, Fronto (in a book now lost), Celsus in his *Book of Truth,* and Porphyry in his lost *Against the Christians.*

39. *Apologia pro Christianis, PG,* 6, 410–11, 426. In col. 426, for example, he says, "Ad imitationem ergo illius Dei Spiritus, qui super aquas ferri dictus est, Proserpinam Jovis filiam vocarunt," with Minerva paralleling the Verbum; the strength of Hercules also anticipates that of Christ.

40. *Ad Autolycum, PG,* 6, 1146–47.

41. *Commentary on the Dream of Scipio* 1.2.13. The text of the *Commentum* has been edited by James Willis (Leipzig: B. G. Teubner, 1963) and translated by William Harris Stahl (1952; rpt. New York and London: Columbia Univ. Press, 1966).

42. *Commentary on the Dream of Scipio* 1.2.9.

43. Peter Dronke, *Fabula: Explorations into the Uses of Myth in Medieval Platonism,* Mittellateinische Studien und Texte, vol. 9 (Leiden and Cologne: E. J. Brill, 1974), p. 176n2, especially in the second book centering on Stoic philosophy (p. 22n).

44. 1.2.10, ed. Dronke, p. 70.

45. 1.2.10–11, ed. Dronke, p. 71.

46. See "The *Ovidius Moralizatus* of Petrus Berchorius: An Introduction and Translation," by William Donald Reynolds (Ph.D. diss., Univ. of Illinois, 1971), p. 46.

47. Much has been written on the history of the four levels and of allegory in the Middle Ages. See especially Caplan, "The Four Senses of Scriptural Interpretation and the Mediaeval Theory of Preaching." In addition to works mentioned in succeeding notes, see also E. C. Knowlton, "Notes on Early Allegory," *JEGP,* 29 (1930), 159–81; Stephen Manning, "The Nun's Priest's Morality and the Medieval Attitude toward Fables," *JEGP,* 59 (1960), 403–16; and D. W. Robertson, Jr., *A Preface to Chaucer: Studies in Medieval Perspective* (1962; rpt. Princeton, N.J.: Princeton Univ. Press, 1969).

48. See Eva Matthews Sanford, "Lucan and the Civil War," *Classical Philology,* 28 (1933), 121–27; and also her "Lucan and his Roman Critics," *Classical Philology,* 26 (1931), 233–47; George Meredith Logan, "Lucan in England: The Influence of the *Pharsalia* on English Letters from the Beginnings through the Sixteenth Century" (Ph.D. diss., Harvard Univ., 1967); David Vessey, *Statius and the Thebaid* (Cambridge: Cambridge Univ. Press, 1973), esp. Chap. Two.

49. The argument for the *Metamorphoses* as an epic work comes from Brooks Otis, *Ovid as an Epic Poet,* 2d ed. (Cambridge: Cambridge Univ. Press, 1970) esp. the first chapter.

50. *Troilus and Criseyde,* 5.1791–92, in *The Complete Works of Geoffrey Chaucer,* ed. Fred C. Robinson, 2d ed. (Boston: Houghton Mifflin, 1957).

51. Henry Osborn Taylor, *The Emergence of Christian Culture in the West: The Classical Heritage of the Middle Ages,* 3d ed. (1911; rpt. New York: Harper, 1958), p. 50.

52. Martianus Capella, *The Marriage of Philology and Mercury* 1.1, trans. William Harris Stahl, Richard Johnson, and E. L. Burge, in vol. 2 of *Martianus*

Capella and the Seven Liberal Arts (New York: Columbia Univ. Press, 1977): "Calliope is glad to have you [Hymen] bless the beginning of her poem concerning the wedding of a god."

53. William Harris Stahl, *The Quadrivium of Martianus Capella: Latin Traditions in the Mathematical Sciences*, vol. 1 of *Martianus Capella and the Seven Liberal Arts* (New York and London: Columbia Univ. Press, 1971), p. 40.

54. The Erfurt Commentator declares: "Sed iste longe nobiliore materia et facundia praecellit, quippe qui nec Tullio in prosa nec Virgilio in metro inferior floruit," cited in Edmund Taite Silk, ed., *Saeculi noni Auctoris in Boetii Consolationem Philosophiae Commentarius*, Papers and Monographs of the American Academy in Rome, vol. 9 (Rome: American Academy, 1935), p. 4.

55. Stahl, *Quadrivium*, p. 56.

56. No. 7, in *Fulgentius the Mythographer*, trans. Leslie George Whitbread (Columbus: Ohio State Univ. Press, 1971).

57. *Inferno* 9, *Purgatorio* 25 and 32. See *Petri Allegherii super Dantis ipsius genitoris Comoediam Commentarium nunc primum in lucem editum*, ed. Vincentio Nannucci (Florence: Guilielmus Piatti, 1845), pp. 124, 471, 523, and 526.

58. Charles Sears Baldwin, *Medieval Rhetoric and Poetic to 1400: Interpreted from Representative Works* (New York: Macmillan, 1928), p. 151: "At the fall of Rome the Trivium was dominated by *rhetorica*; in the Carolingian period by *grammatica*; in the high middle age, by *dialectica*."

59. Gilbert A. A. Grindle, *The Destruction of Paganism in the Roman Empire from Constantine to Justinian* (Oxford: B. H. Blackwell; London: Simpkin, Marshall, Hamilton, Kent, 1892), pp. 1–2, 10; and Walter Woodburn Hyde, *Paganism to Christianity in the Roman Empire* (Philadelphia: Univ. of Pennsylvania Press; London: George Cumberlege, Oxford Univ. Press, 1946), p. 19.

60. H. I. Marrou, *A History of Education in Antiquity*, trans. from the 3d French edition by George Lamb (New York: Sheed and Ward, 1956), pp. 322–23; but see also Chapter Nine on Christianity and classical education.

61. F. J. E. Raby, *A History of Secular Latin Poetry in the Middle Ages*, vol. 1 (Oxford: Clarendon Press, 1934), p. 42.

62. Taylor, *The Emergence of Christian Culture in the West*, p. 49.

63. Raby, p. 99.

64. Marrou, pp. 345–46.

65. Marrou, p. 347; such tensions were witnessed in the careers of Ennodius, Cassiodorus, and St. Gregory the Great.

66. Marrou, p. 347, indicates that the humanistic literary tradition continued in Byzantine strongholds in the north like Ravenna (until 751), and in the far south, Rome and Naples, neither city of which was barbarized; on secular education, see p. 349.

67. Marrou, p. 349.

68. According to Bede, *Historia ecclesiasticia gentis anglorum historia abbatum epistola ac Ecgberctum*, ed. Charles Plummer, 2 vols. (Oxford: Clarendon Press, 1896).

69. Greek had not been much known from the fourth century in Italy, less so in the sixth and seventh centuries, so that "wellnigh" the only knowledge of

Greek was that of the Irish of the seventh century, according to Taylor, *Emergence*, p. 44. For the origin of the Irish priests, see Marrou, p. 349.

70. Ludwig Biehler, *Ireland: Harbinger of the Middle Ages*, trans. from the German 1961 ed. (London, New York, and Toronto: Oxford Univ. Press, 1963), p. 1. But this synthesis was not influenced by that of the Italian scholar Cassiodorus—the Irish monastery had always concentrated on the things of the mind (p. 2).

71. Kathleen O. Elliott and J. P. Elder, "A Critical Edition of the Vatican Mythographers," *Transactions of the American Philological Association*, 78 (1947), 198–99.

72. Marrou, p. 350. Accompanying Alcuin were the Scoti, Clement, Joseph, Dungal; accompanying Paul were Peter of Pisa and Paulinus of Aquileia.

73. When the Scots came to be called Scots, or Scoti, after the Irish emigrants, or Scotti, had colonized the western coast of Caledonia and been converted to Christianity by homeland missionaries, then the Irish were called Hiberni, Hibernienses, Hibernici. John Scottus, or John Scot, was an Irishman, one of the Scotti; Duns Scotus was Scottish, one of the Scoti (Biehler, pp. 2, 4).

74. For example, John Eriugena, 'scion of Eire,' was a theologian and philosopher (Biehler, pp. 10, 4). For other reasons behind the departure of the Irish cleric, see Marrou, pp. 349–50, and also J. M. Clark, *The Abbey of St. Gall as a Centre of Literature and Art* (Cambridge: Cambridge Univ. Press, 1926). On the copying of classics, see Eva Matthews Sanford, "The Use of Classical Latin Authors in the *Libri Manuales*," *Transactions of the American Philological Association*, 55 (1924), 190. Only one *liber manualis* has been found earlier than the ninth century; most of them began to be collected at St. Gall upon Charlemagne's orders, although there were probably earlier collections that linked late classical works with these collections.

75. Biehler, p. 2. The Irish established *schottenmönche*, or monastic houses, in Ratisbon, Vienna, Erfurt, Würzburg, and Nuremberg (p. 4).

76. Irish glosses appeared in early manuscripts in Würzburg, St. Gall, Karlsruhe, and Milan. Apparently it was their scholarship and not their art (except for fine manuscript illuminations) that was preserved (Biehler, p. 4).

77. W. P. Ker, *The Dark Ages* (Edinburgh and London: William Blackwood, 1911), p. 20.

78. The earliest glosses on Martianus, in Welsh, exist in two Cambridge manuscripts (Corpus Christi Mss. 153 and 330); there is also an early commentary on Boethius (Vatican Library Ms. Vat. lat. 3363), possibly by the Welsh Asser. For this attribution see Fabio Troncarelli, *Tradizioni Perdute: L'Antica 'Fortuna' della 'Consolatio Philosophiae'* (Padua: Editrice Antenore, 1980). There is also a very old anonymous commentary on Boethius formerly belonging to St. Gall. The Old High German glosses therein may have been written by one of the many Irish or Scot scholars in exile at the monastery of St. Gall. The manuscript at one time was called St. Gall 845, fols. 3–240, now Naples 845; a longer version exists in Ms. Einsiedeln 179, according to information supplied to me in a letter by Petrus Tax.

79. Baldwin, p. 151.

80. See Wetherbee for the fullest treatment of *Platonism and Poetry in the Twelfth Century.*

81. Raby, p. 307.

82. Raymond Klibansky, "The School of Chartres," in *Twelfth-Century Europe and the Foundations of Modern Society,* eds. Marshall Clagett, Gaines Post, Robert Reynolds (Madison, Milwaukee, and London: Univ. of Wisconsin Press, 1966), p. 5. An example is Bernard of Chartres, with whom this change of organization is associated. He read both pagan and late antique works.

83. Charles Homer Haskins, *The Renaissance of the Twelfth Century* (Cambridge: Harvard Univ. Press, 1927), p. 7. On libraries see James Stuart Beddie, "Libraries in the Twelfth Century: Their Catalogues and Contents," in *Anniversary Essays in Mediaeval History by Students of Charles Homer Haskins* (Boston and New York: Houghton Mifflin, 1929), p. 21. On the popularity of classical texts, see Beddie's "Ancient Classics in the Mediaeval Libraries," *Speculum,* 5 (1930), 3–20. However, the most popular and common book was the Bible (especially Genesis, Kings, Job, the Psalms, Song of Songs, and the Old Testament Apocrypha—Tobit, Judith, Maccabees, Apocalypse, and Paul's Epistles), followed by the Church Fathers, especially Augustine, Ambrose, Jerome, and Gregory (there was neither much Greek nor many Antinicene fathers like Tertullian). Augustine's *De civitate Dei* was his best known work (Beddie, pp. 9–10).

84. For lists of known catalogs compiled to the year 1500, see Beddie, "Libraries," p. 1. The largest libraries in this century, with 342 to 546 volumes, were those of Corbie, Durham, and Cluny (Beddie, "Libraries," p. 2).

85. Beddie, "Ancient Classics," pp. 5, 6, and 10. For a list and history of pagan works used as texts for instruction in the ninth through thirteenth centuries, see M. Boas, "De Librorum Catonianorum Historia atque Compositio," *Mnemosyne,* N.S., 42 (1914), 17–46; and for the twelfth century, see Charles H. Haskins, "A List of Textbooks from the Close of the Twelfth Century," *Harvard Studies in Classical Philology,* 20 (1909), 75–94.

86. See R. W. Hunt, "The Deposit of Latin Classics in the Twelfth-Century Renaissance," in *Classical Influences on European Cultures AD 500–1500,* ed. R. R. Bolgar (Cambridge: Cambridge Univ. Press, 1971), pp. 51–56; and B. L. Ullman, "Classical Authors in Certain Mediaeval *Florilegia*," *Classical Philology,* 27 (1932), 1–42.

87. Haskins, *Renaissance,* p. 6.

88. Edward Kennard Rand, "The Classics in the Thirteenth Century," *Speculum,* 4 (1929), 260.

89. Charles Homer Haskins, *The Renaissance of the Twelfth Century* (Cambridge: Harvard Univ. Press, 1927), p. 93.

90. Rand, "Classics," p. 260.

91. See Sanford, "The Use of Classical Latin Authors," pp. 190–248; Rand, "Classics," p. 264n; and Ullman, "Classical Authors," pp. 1–42.

92. See Beryl Smalley, *English Friars and Antiquity in the Early Fourteenth Century* (Oxford: Basil Blackwell; New York: Barnes and Noble, 1960), and Judson Boyce Allen, "Mythology and the Bible Commentaries and *Moralitates* of

Robert Holkot" (Ph.D. diss., Johns Hopkins Univ., 1963), p. 114.

93. Dante's *Convivio*, trans. William Walrond Jackson (Oxford: Clarendon Press, 1909), p. 100; 3.11, p. 167, from the second book of the *Aeneid*; and 4.23–8, pp. 274–95. See also the edition by Elisa Colesanti, *Opere Minor: La Vita Nuova—Il Convivio—Le Rime*, Il classici azzuri, 30 (Rome: Cremonese, 1956).

94. Dante cites other epic works in discussing the last two stages as well as the *Aeneid* in discussing the first—Old Age is exemplified by Ovid's *Metamorphoses* and the tale of Cephalus in the seventh book, and Decline by Lucan's *Pharsalia* and the life of Marcia.

95. *Rerum Senilium Libri* 4.4, "De quibusdam fictionibus Virgilii," appears in the translation by James Harvey Robinson and Henry Winchester Rolfe, *Petrarch: The First Modern Scholar and Man of Letters*, 2d ed. (New York and London: Putnam and The Knickerbocker Press, 1914), pp. 234–36. For the Latin text consult *Francisci Petrarchae Operum*, vol. 2 (Basel, 1554, rpt. Ridgewood, N.J.: The Gregg Press, 1965), pp. 867–74 (the only complete edition of Petrarch's works, here in facsimile).

96. *Petrarch's Secret or the Soul's Conflict with Passion: Three Dialogues between Himself and S. Augustine*, trans. William H. Draper (London: Chatto and Windus, 1951), p. 82. Subsequent references to the translation will appear in the text. For the Latin text see *De secreto conflictu curarum mearum*, in Francesco Petrarca, *Opere*, ed. Giovanni Ponte (Milan: U. Mursia, 1968), pp. 432–597.

97. "The Ground and Nature of Literary Theory in Bernard Silvester's Twelfth-Century Commentary on the *Aeneid*," trans. Daniel Carl Meerson (Ph.D. diss., Univ. of Chicago, 1967), p. 107 (I.10). Subsequent references to Meerson's translation of the commentary will appear in the text. Where appropriate this will be the second reference appearing in the text, following the primary reference to the Joneses' edition.

98. He may have derived this idea from Macrobius' *Commentum in Somnium Scipionis* 1.10.10 and 1.12.11, wherein the body and its passions are linked with parts of Hades.

99. These tenth-century notes from Bibliothèque Nationale Ms. 7930 (or Parisinus 7930) have been edited by J. J. H. Savage, "Mediaeval Notes on the Sixth *Aeneid* in Parisinus 7930," *Speculum*, 9 (1934), 204–212. The description of the diagram of the circles of Hell occurs on p. 211. Savage believes that this diagram may have influenced Dante's conception in the *Commedia*.

100. The translation of Petrarch's letter to Virgil, "Ad Publium Virgilium Maronem," appears in *Petrarch's Letters to Classical Authors*, trans. Mario Emilio Cosenza (Chicago: Univ. of Chicago Press, 1910), pp. 136–40; the Latin letter from *Familiarium rerum libri*, 24.11, occurs in Francesco Petrarca, *Le Familiari*, vol. 4 of 4 vols., ed. Umberto Bosco, vol. 13 of *Edizione Nazionale delle Opere di Francesco Petrarca* (Florence: G. C. Sansoni, 1942), pp. 251–53. Subsequent references will appear in the text.

101. Research for this essay was accomplished in England during the years 1977–78 under the auspices of a National Endowment for the Humanities Fel-

lowship for Independent Study and in Italy during the years 1980–81 under the auspices of a John Simon Guggenheim Memorial Fellowship. Portions of this essay in different form have been delivered as the following papers: "The Epic Origins of Medieval Mythography" at the Thirteenth Annual Conference on Medieval Studies, The Medieval Institute, Western Michigan University, Kalamazoo, Michigan, on 5 May 1978; and "Moralized Virgil in Dante and Petrarch" at the Fifteenth International Conference on Medieval Studies, The Medieval Institute, Western Michigan University, Kalamazoo, Michigan, on 3 May 1980. I am grateful to Professors Theodore Steinberg and William Reynolds and to Dr. Laura Hodges for reading and commenting on this essay. Any mistakes to be found herein are my own.

THREE

1. Bernard R. Goldstein, *The Arabic Version of Ptolemy's Planetary Hypotheses* (Philadelphia: American Philosophical Society, *Transactions*, 67 [1967], part 4), pp. 3–4.

2. C. S. Lewis, *The Discarded Image. An Introduction to Medieval and Renaissance Literature* (Cambridge: Cambridge Univ. Press, 1964), p. 13.

3. Ibid., p. 10.

4. Ibid., p. 11.

5. Ibid., p. 12.

6. Ibid., p. 97.

7. Arthur O. Lovejoy, *The Great Chain of Being. A Study of the History of an Idea* (Cambridge: Harvard Univ. Press, 1936; New York: Harper & Row, 1960), p. 100.

8. Lewis, pp. 22–91.

9. Aristotle, *On the Heavens*, trans. W. K. C. Guthrie (London: Heinemann; Cambridge: Harvard Univ. Press [Loeb Classical Library], 1939), pp. 253–55 (298a).

10. Aristarchus of Samos, *On the Sizes and Distances of the Sun and Moon*, trans. T. L. Heath, in *Aristarchus of Samos the Ancient Copernicus* (Oxford: Clarendon Press, 1913), pp. 351–411.

11. Otto Neugebauer, *A History of Ancient Mathematical Astronomy*, 3 parts (New York, Heidelberg, Berlin: Springer, 1975), pp. 652–54.

12. Noel M. Swerdlow, "Hipparchus on the Distance of the Sun," *Centaurus*, 14 (1969), 287–305; Gerald J. Toomer, "Hipparchus on the Distances of the Sun and Moon," *Archive for History of Exact Sciences*, 14 (1975), 126–42. Hipparchus' method was as follows.

In the diagram (not drawn to scale), ND is the distance from the earth to the sun and $NH = NP$ is the distance from the earth to the moon. Hipparchus stated that at the moon's mean distance its apparent diameter is equal to the sun's and equal to $1/650$ part of its circle. He further assumed that the sun's parallax angle (angle CMN) was at most 7' of arc, which corresponds to a distance of 490 e.r., or $ND = 490NM$. Finally, from timing lunar eclipses, he determined

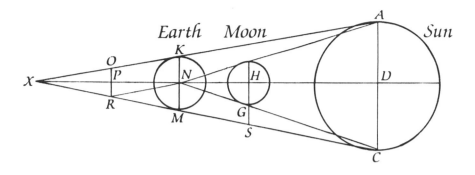

that the width of the earth's shadow cone at the moon's distance is 2½ times the lunar diameter, or $PR = 2HG$. From the geometry of the diagram he could obtain the relationship:

$$\mathrm{NH} = \frac{\mathrm{ND}}{\frac{7}{1300}\,\mathrm{ND} - 1}$$

in which the earth's radius, *NM*, is equal to unity. Knowing that $ND = 490$ e.r., Hipparchus calculated the moon's distance, *NH*, to be 67⅓ e.r.

Ptolemy reversed Hipparchus' procedure. Having determined the moon's distance by other methods, he used the diagram to calculate the distance of the sun.

13. Ptolemy, *Almagest*, Book V, Chapters 11–16; trans. Gerald G. Toomer, *Ptolemy's Almagest* (London: Duckworth, 1984) pp. 243–57.

14. Goldstein, *Arabic Version*, pp. 3–4.

15. Ibid., pp. 7–8.

16. Ibid., pp. 8–9.

17. Thus, while Ptolemy calculated the sun's (mean) distance to be 1210 e.r., the modern figure is about 23,250 e.r.

18. Al-Farghānī, *Differentie Scientie Astrorum*, Ch. 21, 22, trans. John of Seville, ed. Francis J. Carmody (Berkeley, privately printed, 1943), pp. 38–40.

19. In Oxford alone, Paget Toynbee found no fewer than twenty manuscript copies of Al-Farghānī's book. See Toynbee, "Dante's Obligation to Alfraganus in the *Vita Nuova* and *Convivio*," Romania 24 (1895), pp. 413–15.

20. Gerald J. Toomer and Francis S. Benjamin, ed. and trans., *Campanus of Novara and Medieval Planetary Theory* (Madison: Univ. of Wisconsin Press, 1971), pp. 358–63.

21. Lovejoy, pp. 99–101; E. M. W. Tillyard, *The Elizabethan World Picture* (London: Chatto & Windus, 1943; New York: Random House, 1959), pp. 39–40.

22. John of Seville translated *Jawāmiᶜ* in 1137 under the title *Differentie Scientie Astrorum*. Gerard of Cremona translated it in 1175 and gave it the title *Liber de Aggregationibus Scientie Stellarum at Principiis Celestium Motarum*.

John's translation was printed in numerous editions beginning in 1493. The title *Elementa Astronomica* was introduced with the edition of Jacob Christmann (Frankfurt, 1590) and used by Jacob Golius in his retranslation from the Arabic (Amsterdam, 1669). Gerard's translation was not printed until 1910.

23. Lynn Thorndike, *The Sphere of Sacrobosco and its Commentators* (Chicago: Univ. of Chicago Press, 1949), pp. 195, 243.

24. See n. 20.

25. *Theorica Planetarum Gerardi*, ed. Francis J. Carmody (Berkeley: privately printed, 1942).

26. *The Astronomical Works of Thabit B. Qurra*, ed. Francis J. Carmody (Berkeley: Univ. of California Press, 1960), pp. 145–48.

27. Francis J. Carmody, "Leopold of Austria 'Li Compilacions de le Science des Estoilles,' Books I–III. Edited from Ms. French 613 of the Bibliothèque Nationale, with Notes and Glossary," *University of California Publications in Modern Philology*, 33, part 2 (1947), p. 69: "The sizes of the celestial bodies are fourteen. The sun has 166 1/32 times the quantity of the earth; the largest stars 115 times; Jupiter 95 times; Saturn 91 times, those that are of the second magnitude of the fixed stars 90 times; those that are of the third magnitude 70 times; of the fourth 50 times; of the fifth 36 times; of the sixth 18 times; Mars 1 1/16 times; Venus with regard to the Earth the 37th part; the Moon the 39th [part] and a bit more; Mercury one part of 22,000 of the Earth."

28. Ibid.

29. Roger Bacon, *Opus Maius*, trans. Robert Belle Burke, 2 vols. (Philadelphia: Univ. of Pennsylvania Press, 1928), I, pp. 249–50.

30. Ibid., p. 250.

31. Ibid., pp. 250–51.

32. Ibid., pp. 258–59.

33. Ibid., p. 258.

34. *William Caxton's Mirror of the World*, ed. O. H. Prior, *Early English Text Society*, 1913, extra series, no. 110, pp. 170–71.

35. Ibid., pp. 171–72.

36. Ibid., p. 172.

37. Ibid., pp. 125–169.

38. Orosius, *Historiarum adversus Paganos Libri VII*, ed. C. Zangmeister (Vienna, 1882; New York: Johnson Reprint, 1966), pp. 6, 564.

39. C. D'Evelyn and A. J. Mill, eds., *The South English Legendary*, 3 vols., *Early English Text Society*, nos. 235, 236 (1956), 244 (1959), II, p. 415.

40. Ibid., pp. 417–18.

41. Al-Farghānī, *Differentie*, pp. 13–14.

42. Moses Maimonides, *The Guide of the Perplexed*, trans. Shlomo Pines (Chicago: Univ. of Chicago Press, 1969), p. 456.

43. Bacon, *Opus Maius*, I, p. 247.

44. *Caxton's Mirror*, p. 171.

45. Dante Alighieri, *Convivio*, ii, 7, 8, 14. See M. A. Orr, *Dante and the Early Astronomers* (London, 1913; rpt. London: Allan Wingate, 1966), p. 309; and Toynbee, "Dante's Obligation," pp. 422–32.

46. Dante Alighieri, *The Divine Comedy*, trans. C. S. Singleton, 3 vols. in 6 parts (Princeton, N.J.: Princeton Univ. Press, 1970–75), III, *Paradiso*, part 1, p. 103.

47. *Ptolemy's Almagest*, pp. 256–57 (Bk. v, ch. 15); Al-Farghānī, *Differentie*, p. 38 (ch. 21). See also Orr, *Dante and the Early Astronomers*, pp. 310–11.

48. Material in this paper is drawn from my book *Measuring the Universe: Cosmic Dimensions from Aristarchus to Halley* (Chicago: Univ. of Chicago Press, 1985).

FOUR

1. Quotations from Chaucer are from *The Works of Geoffrey Chaucer*, 2d ed., F. N. Robinson (Boston, Mass.: Houghton Mifflin, 1957).

2. D. W. Robertson, Jr., *A Preface to Chaucer* (Princeton, N.J.: Princeton Univ. Press, 1962), p. 243.

3. *Der Trojaroman des Benoît de Sainte-Maure*, ed. Kurt Reichenberger (Tübingen: M. Niemeyer, 1963), pp. 29–30.

4. For the Latin texts and commentary, see Edmond Faral, *Les arts poétiques du XIIe et du XIIIe siècle* (Paris: É. Champion, 1924), pp. 75–81, 118–32, 214–17, 310–12. For an excellent discussion of these and other influences on Chaucer's portraits and for references to other studies, see Jill Mann, "The 'General Prologue' and the 'Descriptio' Tradition," in *Chaucer and Medieval Estates Satire* (Cambridge: Cambridge Univ. Press, 1973), pp. 176–86.

5. Faral, p. 76.

6. Aristotle, *Physiognomia*, in *Minor Works*, trans. W. S. Hett (Loeb Classics, London: William Heinemann, 1936), I, 121–23.

7. Suetonius, *The Lives of the Caesars*, trans. J. C. Rolfe (Loeb Classics, London: William Heinemann, 1928), I, 245.

8. Edith C. Evans, "Physiognomics in the Ancient World," *Transactions of the American Philosophical Society*, N.S., 59, part 5 (1969), 54.

9. Alice M. Colby, *The Portrait in Twelfth-Century French Literature* (Geneva, 1965), pp. 49–50. For noses in Middle English, see Walter Clyde Curry, *The Middle English Ideal of Personal Beauty; as Found in the Metrical Romances, Chronicles, and Legends of the XIII, XIV, and XV Centuries* (Baltimore: J. H. Furst Co., 1916), pp. 63–65.

10. Colby, pp. 72–88.

11. Lines 286–311.

12. *Aucassin and Nicolette*, trans. Pauline Matarasso (Harmondsworth, Middlesex, and Baltimore: Penguin Books, 1971). "Grans estoit et mervellex et lais et hidex; il avoit une grande hure plus noire q'une carbouclee, et avoit plus de planne paume entre deus ex, et avoit unes grandes joes et un grandisme nes plat et unes grans narines lees et unes grosses levres plus rouges d'une carbounee et uns grans dens gaunes et lais" (XXIV.13–19), *Aucassin et Nicolette*, ed. Mario Roques (Paris: Honoré Champion, 1965).

13. Quoted from the Middle English *Romaunt of the Rose* in Robinson. The

noses belong, in the order cited, to Ydelnesse (545), Myrthe (823), Beaute (1023–24), Fraunchise (1215–16), and Gladnesse (865–66). In the French (ed. Ernest Langlois, SATF [Paris, 1914–24]) the noses are described as "bien fait a droiture" (Oiseuse, 532), "fait par grant entente" (Déduit, 808), "bien fait" (Beauté, 1009), "n'ot pas nés orlenois, Ançois ot nés lonc et traitiz" (Franchise, 1194–95), and "L'en nou feïst pas miauz de cire" (Liesse, 850).

14. The French reads:

> *Granz fu e noirs e hericiez,*
> *S'ot les iauz roges come feus,*
> *Le nés fronciés, le vis hisdeus. (2922–24)*

15. *Romaunt*, 4029–30. The French reads, "vos seiez fel e estouz, Pleins de rampones e d'outrage" (3698–99).

16. Giovan Battista della Porta, *Della fisonomia dell'uomo*, ed. Mario Cicognani (Milan: Longanesi and C., 1971), p. 251.

17. Walter Clyde Curry, *Chaucer and the Mediaeval Sciences*, 2d ed. (New York: Barnes & Noble, 1960), pp. 85–90.

18. Ibid., p. 87. Curry is quoting Richard Saunders, *Physiognomie, and Chiromancie, Metoposcopy, Dreams, and the Art of Memory* (London, 1671).

19. Ibid., p. 90.

20. Donald R. Howard, *The Idea of the Canterbury Tales* (Berkeley, Calif.: Univ. of California Press, 1976), pp. 90–92.

21. Thomas Bodkin, *The Wilton Diptych*, Gallery Books No. 16 (London: Percy Lund Humphries and Co., n.d.), pp. 6–7.

22. John Pope-Hennessy, *The Portrait in the Renaissance* (New York: Pantheon, 1966), passim. For other notable noses, see especially Piero della Francesca's portrait of Federigo da Montefeltro (p. 160), Pontormo's of Cosimo de'Medici il Vecchio (p. 180), and Holbein the Younger's of Henry VIII (p. 193).

23. Della Porta, pp. 252–53.

24. Laurence Sterne, *Tristram Shandy*, Book III, Chapter 33.

25. George L. Mosse, *Toward the Final Solution: A History of European Racism* (New York: Harper & Row, 1978), p. 29. On the esthetic basis of the new science of anthropology, see Chapter Two, "From Science to Art: The Birth of Stereotypes," pp. 17–34.

26. Johann Casper Lavater, *Physignomische Fragmente, zur Beförderung der Menschenkentniss und Menschenliebe*, 4 vols., facsimile of the first Leipzig ed. (Zürich: Orell Füssli, 1968–69). Translations from Lavater are my own.

27. Graeme Tytler, *Physiognomy in the European Novel: Faces and Fortunes* (Princeton, N.J.: Princeton Univ. Press, 1982), p. 63.

28. Ibid., p. 72.

29. Ibid., p. 74.

30. *Oliver Twist*, Chapter 8.

31. *Ivanhoe*, Chapter 8.

32. Ibid., Chapter 5.

33. Georg Lukács, *The Historical Novel*, trans. Hannah Mitchell and Stanley Mitchell (London: Merlin Press, 1962), pp. 30–63.

34. Mosse, plate 7.

35. *Tristram Shandy*, Book III, Chapter 41.

FIVE

1. Antonio-Schinella Conti, *Prose e poesi*, 2 vols. (Venice: Giambattista Pasquali, 1739–56), 2, p. 26.

2. Roger North, *The Lives of the Norths*, ed. Augustus Jessopp, 3 vols. (London: Bell, 1890), pp. 3, 15.

3. *The Philosophical Works of Descartes*, trans. Elizabeth S. Haldane and G. R. T. Ross, 2 vols. (New York: Dover, 1955), 1, p. 191.

4. Galileo, *Dialogue Concerning the Two Chief World Systems*, trans. Stillman Drake (Berkeley: Univ. of California Press, 1962), p. 462.

5. See their correspondence after 1687, in which the *Principia* figured very prominently, in Christiaan Huygens, *Oeuvres complètes*, 22 vols. (La Haye: Nijhoff, 1888–1950), 9 and 10, passim.

6. *Oeuvres complètes de Molière*, ed. Gustave Michaut, 11 vols. (Paris: Richelieu, 1947–49), 10, p. 186.

7. *Philosophical Works*, 1, p. 47.

8. *Oeuvres de Descartes*, eds. Charles Adam and Paul Tannery, 10 vols. (Paris: L. Cerf, 1897–1913), 8, pp. 144–56, 275–311.

9. Cambridge University Library, Add. MS. 3970.3, ff. 473–74.

10. Add. MS. 3996, f. 102.

11. *Principles of Philosophy* (1644); *Oeuvres Descartes* 8, pp. 212–14.

12. Add. MS. 3996, f. 121^v.

13. *The Correspondence of Isaac Newton*, eds. H. W. Turnbull, J. F. Scott, A. R. Hall, and Laura Tilling, 7 vols. (Cambridge: Cambridge Univ. Press, 1959–77), 1, pp. 362–85.

14. Fatio to Huygens, 14 June 1687; Huygens to Fatio, 1 July 1687; *Oeuvres Huygens*, 9, pp. 168–69, 190–91.

15. *Principia*, trans. Motte-Cajori (Berkeley: Univ. of California Press, 1934), pp. 164, 192.

16. *Opticks* (New York: Dover, 1952), pp. 350–52.

17. Cotes to Newton, 18 March 1713; *Correspondence*, 5, p. 392.

18. Bodleian Library, MS. Don. b. 15.

19. Add. MS. 3975, p. 51.

20. Notebook in the Fitzwilliam Museum, Cambridge, n.p. Notes on Basil Valentine, King's College, Cambridge, Keynes MS. 64. Notes on Sendivogius, Keynes MS. 19. Notes on Philalethes, Keynes MSS. 51 and 52. Notes on Maier, Keynes MS. 29. (Keynes MS. 29 contains the material that Newton used in his letter of 18 May 1669 to Francis Aston; *Correspondence*, 1, p. 11.) MS. Var. 259 in the Jewish National and University Library, Jerusalem, also contains, along with some other alchemical papers, early notes on Artephius, Flammel, Sendivogius, d'Espagnet, Augurello, Philalethes, Hermes (*Tabula smaragdina*), and several pieces in the *Theatrum chemicum*.

21. Add. MS. 3975, pp. 80–83.

22. Keynes MS. 32.

23. All of the manuscripts of the *Index chemicus* except the drafts between the second and third versions are in Keynes MS. 30. The drafts are in a manuscript in the Yale Medical Library.

24. John Harrison, *The Library of Isaac Newton* (Cambridge: Cambridge Univ. Press, 1978), p. 8. By my count the proportion of alchemical works was higher than Harrison's figure of 9.5 percent.

25. Keynes MS. 67. Newton's paragraph is on f. 68ᵛ. Notes and copies from the collection are in Keynes MS. 62.

26. Keynes MS. 22, 24, 33, 51, 52, 65. A manuscript in the Yale Medical Library. Meheux to Newton, 2 March 1683; *Correspondence* 2, p. 386. Newton's memorandum on the visit in 1696 is printed in *Correspondence* 4, pp. 196–98; the original is in Keynes MS. 26; another version is in the Joseph Halle Schaffner Collection, University of Chicago Library.

27. Dibner Library, Smithsonian Institution, Burndy MS. 16. Keynes MS. 18. (The text and translation of *The Key* are published in B. J. T. Dobbs, *The Foundations of Newton's Alchemy* [Cambridge: Cambridge Univ. Press, 1975], pp. 251–55. All of Mrs. Dobbs' book contributes to its explication.) Burndy MS. 10. Keynes MS. 28.

28. The sheet, now with Keynes MS. 30, belongs with Keynes MS. 35, which contains the drafts of chapters.

29. Keynes MS. 49.

30. Keynes MS. 48.

31. Keynes MSS. 21 and 53. Babson College, Babson MS. 420.

32. Drafts are found in Keynes MSS. 40 and 41, Babson MS. 417, and Burndy MS. 17.

33. Keynes MS. 135.

34. Add. MS. 3973, f. 17. Add. MS. 3975, p. 149.

35. Add. MS. 3973, f. 29.

36. David Brewster, *Memoirs of the Life, Writings, and Discoveries of Sir Isaac Newton*, 2d ed., 2 vols. (Edinburgh: Edmondston and Douglas, 1860), 2, pp. 300–302.

37. Eirenaeus Philalethes, "An Exposition upon Sir George Ripley's Preface," in *Ripley Reviv'd* (London: W. Cooper, 1678), p. 28. Newton's repetition of this passage is found in Keynes MS. 30; Keynes MS. 34, f. 1; Keynes MS. 35, sheet 4; Keynes MS. 48, ff. 16–16ᵛ; Keynes MS. 51, f. 1ᵛ; Babson MS. 420, p. 8.

38. Add. MS. 3973, ff. 13, 21, 42. Add. MS. 3975, pp. 104–105, 108–109, 281.

39. Burndy MS. 16.

40. *Correspondence*, 1, pp. 362–85.

41. John Herivel, *The Background to Newton's Principia* (Oxford: Oxford Univ. Press, 1965), pp. 257–74; translation, pp. 277–89.

42. *Principia*, p. xviii.

43. A. R. Hall and M. B. Hall, eds., *Unpublished Scientific Papers of Isaac Newton* (Cambridge: Cambridge Univ. Press, 1962), p. 333. Newton later redrafted this material for his preface (ibid., pp. 302–308), then finally suppressed it. The two essays constitute first drafts of his later Query 31.

44. Ibid., pp. 333–35.

45. *Opticks*, pp. 397–400.

46. *Principia*, pp. 226–28. *Opticks*, p. 272.

47. Ibid., p. 275.

48. *Isaac Newton's Papers & Letters on Natural Philosophy*, ed. I. Bernard Cohen (Cambridge, Mass.: Harvard Univ. Press, 1958), pp. 257–58.

49. "Conclusio," *Unpublished Papers*, p. 341. Cf. draft preface, ibid., p.303; a draft from the early 90s for a revision of Book III, ibid., p. 317, and Add. MS. 3965.6, f. 266ᵛ.

50. See Newton's notes in Keynes MS. 19, f. 1.

51. See especially J. E. McGuire, "Force, Active Principles, and Newton's Invisible Realm," *Ambix*, 15 (1968), 154–208.

52. A date in the spring of 1686 appears in his record of experiments, apparently marking the beginning of a new series. Add. MS. 3975, p. 150.

53. Babson MS. 420, p. 18ᵃ. This passage occurs in a draft. The final version (p. 17) watered it down a bit without retracting its central assertion.

54. Keynes MS. 13, f. 1ᵛ bis; Keynes MS. 56, f. 1. Public Record Office, *Mint Papers* 19.5, ff. 42, 54ᵛ.

55. Harrison, p. 9.

56. Ibid., items 1138, 1302 and 1644. See Yworth to Newton, c.1702; *Correspondence*, 7, p. 441.

SIX

1. Derek W. Forrest, *Francis Galton: The Life and Work of a Victorian Genius* (New York: Taplinger, 1974).

2. Charles Darwin, *On the Origin of Species by Means of Natural Selection, or the Preservation of Favored Races in the Struggle for Life* (London: Murray, 1859).

3. Charles Darwin, *The Descent of Man* (London: Murray, 1871).

4. Francis Galton, *Hereditary Genius: an Inquiry into its Laws and Consequences* (London: Macmillan, 1869).

5. Ibid., p. 276.

6. Stephen Jay Gould, *The Mismeasure of Man* (New York: Norton, 1981).

7. Thomas S. Kuhn, *The Structure of Scientific Revolutions* (Chicago: Univ. of Chicago Press, 1962).

8. See Howard Resnikoff's essay, this volume.

9. Abraham Pais, *Subtle is the Lord: The Science and the Life of Albert Einstein* (New York: Oxford Univ. Press, 1982).

10. Ibid., p. 443.

11. Ibid., p. 79.

12. Kuhn, p. 137.

13. Ernst Mayr, *The Growth of Biological Thought: Diversity, Evolution, and Inheritance* (Cambridge, Mass.: Belknap, 1982).

14. Gregor Mendel, "Versuche ueber Pflanzen-hybriden," *Verh. Natur. Vereins Bruenn* 4 (1865), 3–57.

15. Peter L. Brent, *Charles Darwin, a Man of Enlarged Curiosity* (New York: Harper and Row, 1981).

16. Mayr, p. 566.

17. Mayr, p. 45.

18. Francis Galton, *English Men of Science, Their Nature and Nurture* (London: Macmillan, 1874).

19. Cyril L. Burt, *The Gifted Child* (London: Hodder and Staughton, 1975).

20. David Cohen, *J. B. Watson, the Founder of Behaviorism: A Biography* (Boston: Routledge and Kegan Paul, 1979).

21. Edward O. Wilson, *Sociobiology* (Cambridge, Mass.: Harvard Univ. Press, 1975).

22. R. D. Laing, *Knots* (New York: Pantheon, 1970).

23. George E. Marcus, Review of *Margaret Mead and Samoa*, in *New York Times Book Review* (March 27, 1983).

24. Lee J. Cronbach, "The Two Disciplines of Scientific Psychology," *The American Psychologist*, 12, 671.

25. Kenneth E. Bock, *Human Nature and History* (New York: Columbia Univ. Press, 1980).

26. Arthur R. Jensen, *Straight Talk About Mental Tests* (New York: Free Press, 1981).

27. Hans J. Eysenck, *The Inequality of Man* (San Diego: R. R. Knapp, 1973).

28. Cohen, *J. B. Watson*.

29. Leslie Spencer Hearnshaw, *Cyril Burt, Psychologist* (Ithaca, N.Y.: Cornell Univ. Press, 1979).

ABOUT THE AUTHORS

Jane Chance is Professor of English at Rice University, where she teaches medieval literature. Her books include *The Genius Figure in Antiquity and the Middle Ages, Tolkien's Art: "A Mythology for England," Woman as Hero in Old English Literature*, and another co-edited book, *Approaches to Teaching Sir Gawain and the Green Knight*. She is now at work on a two-volume study of *The Mythographic Tradition in the Middle Ages*.

Alfred David is Professor of English at Indiana University, where he teaches medieval literature. He is the author of *The Strumpet Muse: Art and Morals in Chaucer's Poetry* and the co-editor of the *Minor Poems* for the *Chaucer Variorum*.

Howard L. Resnikoff is a founder, Vice President, and Director of Research at Thinking Machines Corporation. He is currently teaching at Harvard University, where he was previously Associate Vice President. He is the author of numerous books, including *Mathematics in Civilization* with R. O. Wells, Jr.

George R. Terrell received his Ph.D. in Mathematics from Rice University in 1978; he now teaches in the Mathematical Sciences Department. He was led to his primary area of research, mathematical statistics, by a longstanding interest in the behavioral sciences.

Albert Van Helden, Professor of history at Rice University, teaches the histories of science, technology, and medicine. His book *Measuring the Universe: Cosmic Dimensions from Aristarchus to Halley* was published by the University of Chicago Press in the spring of 1985.

R. O. Wells, Jr., is Professor of mathematics at Rice University. His areas of research include algebraic geometry, several complex variables, and mathematical physics. He is the author of *Differential Analysis on Complex Manifolds* and, with Howard Resnikoff, *Mathematics and Civilization*.

Richard S. Westfall, Professor of history of science at Indiana University, is a student of seventeenth-century science and the author of *Never at Rest, a Biography of Isaac Newton*.

INDEX OF AUTHORS
AND INVENTORS DISCUSSED